CREATORS 2018-2020

文化實驗三年索引

目次 CONTENT

序 FOREWORD

計畫 PROJECTS

關於文化實驗 ABOUT CULTURAL EXPERIMENTATION

附錄 APPENDIX

序

文__謝翠玉

財團法人
臺灣生活美學基金會 執行長

臺灣當代文化實驗場（C-LAB）自2018年啟動以來，以一個面向未來的新型態藝文機構為目標，致力作為孕育文化藝術生產前端的創新基地。在藝術、科技、社會三者交融的視野之下，持續推動實驗性、具創新思維的文化實踐，並透過多樣化的跨領域共製、技術媒合與異業合作，探索文化實驗的未來式。

創造活動在最初期的發想、探索與實驗，往往是藝術文化生產在步入產業化之後，最容易被忽略的階段。為了打造一個有機且具有包容力的支持系統，C-LAB在成立之初即推出「CREATORS創作／研發支持計畫」，此計畫依「未來視野、實驗精神、人文脈絡、跨域共融」四項指標，每年遴選來自不同領域團隊的文化實驗提案。在C-LAB提供的空間、經費、技術媒合與行政資源挹注之下，團隊將以六至九個月的時間發展實現各自的專案。

與國內現行類似藝文補助機制相較，CREATORS計畫試圖更朝向開放性與實驗性，專注於扶植計畫前期的研究開發，其不以傳統認定之作品、展覽、映演製作等為唯一的預期成果想像。在「從0到1」的過程中，CREATORS計畫鼓勵創作或文化生產「從發想到實踐第一步」的最初階段，也逐漸形塑此計畫成為臺灣藝術文化創作的發電機。

作為C-LAB發展的核心項目之一，CREATORS計畫從推出至今已累計扶植超過50組團隊參與，其所推進的各項文化實驗提案，後續也在C-LAB及國內外各大藝文機構所推出的各項展覽、藝術節與公共節目中開花結果成為成熟的製作；而每年針對各項計畫的觀察員書寫，則累積出可貴的知識生產，為這些當初仍在發展中的計畫，留下了對於過程意義的深刻反思。

這本專書關於 CREATORS 計畫初始三年的階段性面貌，書中所收錄的各項計畫以多樣的主題、形式與關懷，彰顯了 C-LAB 作為一個文化實驗基地的核心精神。透過出版回顧過往的同時，新一年度的 CREATORS 計畫正熱切進行中，以一種動態而有機的方式，描繪著我們對於文化實驗的想像。

Foreword

TEXT_HSIEH Tsui-Yu

Executive Director,
Taiwan Living Arts Foundation

Launched in 2018, the Taiwan Contemporary Culture Lab (C-LAB) has endeavored to become a visionary and innovative cultural and art institution that facilitates cultural and art production ventures. Integrating the perspectives of art, technology, and society, C-LAB seeks to promote experimental cultural practices with inventive ideas and to explore the future of cultural experimentation through collaborative interdisciplinary productions, technological coordination, and cross-industry cooperation.

The experimentation, conceptualization, and exploration of creative activities are frequently undervalued once cultural and art production have been industrialized. To develop an organic and inclusive support system, the CREATORS Creation/Research Support Program was established since the very beginning of C-LAB to foster experimental cultural proposals from various fields, which are selected each year according to four criteria: prospectivity, experimentality, cultural contextuality, and interdisciplinarity. With C-LAB's spaces, subsidies, and technological and administrative assistance, the individuals selected for the program have six to nine months to develop their projects and turn their ideas into reality.

Compared with similar domestic cultural and art subsidy systems, CREATORS

Program takes a more open and experimental route that focuses on endorsing early-stage research and development rather than the final outcomes of artworks, exhibitions, performances, or cultural productions, such as are expected by traditional subsidy logics. In this process of “from 0 to 1,” CREATORS Program supports the creators’ preliminary efforts to embody their ideas, making this program a generator of art and cultural production in Taiwan.

As one of the core programs of C-LAB, CREATORS Program has supported more than fifty cultural experiment projects, many of which have evolved into mature productions and been part of various exhibitions, art festivals, and public programs, both domestically and abroad. In the meantime, the reviewers’ writings of these projects each year have generated invaluable knowledge and represent profound reflections on the process and progress of the projects in their development phase.

This book is dedicated to documenting the progress of the CREATORS program over the past three years—since its very beginning—and the projects included in this book bear testimony to how C-LAB endeavors to be a base of cultural experimentation, with diverse subjects, formats, and concerns. While looking back, CREATORS Program is also in full swing for next year, illustrating further visions of cultural experimentation with its dynamic and organic proposition.

前言：從工作室到一個創造實驗的聚落

文__游崴

C-LAB 臺灣當代文化實驗場自 2018 啟動以來，「CREATORS 創作／研發支持計畫」（簡稱 CREATORS 計畫）即是一項核心項目。這個混合了補助、進駐、育成陪伴、公眾活動及潛在的展演及委託製作的支持系統，瞄準了藝術文化創造活動初始的研究開發階段，每年透過公開徵選，挑選具有實驗性、創新精神、跨領域向度的計畫，使其在 C-LAB 資源挹注下，進行為期 6 至 9 個月的發展實踐。

CREATORS 中的各項計畫執行者，自然是契約關係中的獲補者，但在另一個比較不枯燥的版本中，他們更是 C-LAB 有力的合作者，共同為這塊基地生產內容、提出想像，並以滾動的方式形塑了文化實驗的意義。

CREATORS 計畫從一開始單純的空間進駐取向，逐年進行細部調整，朝向一個複合式的支持機制，以扣合當代藝術文化生產的需求。首先，在 2019 年起在「創研進駐」之外，新增不帶空間需求的「創研支持」模式，讓 CREATORS 計畫能與更多分散在臺灣各處在地深耕的團隊彼此串連。2020 年則新增「CREATORS 聲鬥陣：創作新秀進駐計畫」，聚焦於環繞聲場創作為導向，拔擢實驗聲響與音樂的創作新秀，並以與 C-LAB 臺灣聲響實驗室合作共製為目標。

CREATORS 在實驗什麼？

相較於傳統定義下慣以展演映製作為最終產物的「成果論」，CREATORS 計畫更看重藝術文化生產過程中的行動意義，以及潛在的公共性。如何從「產物」到「生產過程」的焦點轉移中梳理觀點，進而對計畫發展產生有意義的反饋？是 CREATORS 計畫這幾年發展歷程中一個基礎的問題意識。據此，CREATORS 計畫自創立之初，即規畫有觀察員制度：每項計畫都依領域分配有一名「陪伴觀察員」提供諮詢及討論的機會，讓團隊不用在計畫發展的道路上獨自摸索；2019 年起則新增以觀察書寫為導向的「年度觀察員」，並在隔年進一步擴充為「年度觀察團」，連結《CLABO 實驗波》的專輯企畫，透過更有動能的企畫精神，讓這些「針對計畫過程的書寫」，能在既有的藝評及報導書寫文類中，探索出一條新的路徑。

CREATORS 計畫對於過程中的行動意義、公共性及跨領域的視野，很大程度來延伸自於過往的藝術對於「文化生產」在觀念與形式上的探索及反思，而當代藝術如今所開拓出的疆界則是重要參照之一。如今，在宏觀的文化場域中，當代藝術仍是一個界線相對寬鬆、最能匯集想像力的子場域，也是一個關於跨領域的理想實驗場。藝術家透過造形的、感知的、觀念的藝術製作，有些延伸出對於社會、群眾的連結，有些則致力於在不同學科領域、知識系統之間創造火花。

在 CREATORS 計畫中，佔大多數的「藝術家」角色，往往不只是傳統概念下的創造者，更像是這些外延出去的多樣文化實踐之行動者或代理人（agents），帶領群眾進行田野踏查、主持工作坊，或透過講座分享研究成果。更有不少 CREATORS 計畫的獲補者，本身就是文化生產鏈的中介者角色，像是編輯、評論者、策展人、文化行動者、非營利團隊，及其它自我組織者等。

在 2018 至 2020 年間獲選的 CREAOTRS 計畫中，主題與內容多樣，但可觀察到幾個面向。首先是對於藝術的語言技藝實驗，像是從當代馬戲、南管、越南籌歌（Ca trù）、弄鐃、起乩中開發表演藝術中的身體語言；或是從動態影像、記譜學、推測設計、現成物拼接中開拓新的視覺形式。其次，是對新媒體技術在藝術創作上的應用開發，如點雲（point cloud）掃瞄、VR 虛擬實境、即時音像、環境數據偵測、波場合成（Wave Field Synthesis, WFS）、基因改造、演算法、網路爬蟲、Ambisonics 環繞聲場、人工智能之人機共創等。

此外，也有許多帶跨領域精神的不同主題之文史調研，如戰後外省人離散經驗、臺灣同志污名史、火山與地理 / 地質學、臺灣糖業史、礦工文化、日治時期歌謠等，以及從媒體批判、網路文化、城市空間出發的研究。另外，也有一些團隊則著力於自我組織與文化行動，如創作者社群平台、聚落保存、閒置空間再利用、雜草採集、發酵與城市空間、都市園藝等。這些計畫用各自的方式跨出學科領域的象牙塔，以不同路徑連結外在現實，有時更緊扣著臺灣當代社會的脈動，處理資訊戰、香港反送中運動、COVID-19 疫情等議題。

當我們思考工作室

從藝術的角度思考文化生產的過程意義，很大程度意味著對「工作室」的歷史、形式與意義進行反思。在那裡，（藝術家的）工作室是一種美學的、政治的、社會的建制。早期的「工作室」可以是一個包藏了藝術家創造活動終極意義的私密場所、一個堆積創作素材與未完成品的倉庫，有時它也是一個混合了展示、交流與派對的空間。但大多時候這些工作室定義仍建立在藝術物件的生產前題。

1960-70 年代的觀念藝術家在激進地改寫了「作品」定義的同時，也進一步挑戰了對「工作室」的傳統想像。在所謂「後工作室」（post-studio）時代下，有些藝術家的工作室更像是辦公室與工廠，裡面雇有不只一位行政助理或是實作的勞工；有些則更像是一個社交網絡下的節點、一個不斷與外部發生各式關係的社會場所；有些則不再需要實體空間，靠一台筆記型電腦就定義了所謂工作室，地點則可能在臥室、咖啡館或旅館大廳。「藝術家進駐」（artist-in-residency）機制，週期性地提供藝術家旅行並在異地進行短期創作與研究的機會。在這種暫時性的工作室中，創造活動混合了全球化時代下的移動經驗，並更帶有更多旨在拓展人際網絡的社交意義。

我們試著以藝術生產對於「工作室」的歷史想像及當代藝術的實踐為基礎，並混合新創的孵化器（incubator）模式，引入計畫陪伴、育成培力等更主動的數計，思考 CREATORS 這個以「工作進駐」為核心的計畫的可能模樣：它將保留一部分現代主義式的、作為創造活動之私密空間的功能，但朝向一個讓創造活動可以開放運作的場域、一個具備公共性的準社會空間。

基於此種想法，CREATORS 的獲選團隊在計畫期程間，將透過不同形式的公眾活動，使計畫能在發展過程中持續與外部發生關係。我們期望這些活動不僅僅是在公共關係的層面上為了「讓實驗可以被看見」而產生的某種教育推廣或宣傳事務，而是能更深刻、有機地融入計畫內容之中，讓發展歷程的外部連結能適當地產生回饋，變成讓計畫向前推進或校準方向的動能之一。在我們眼中，這些計畫發展過程中延伸出的各式公眾活動意義非凡，其重要性不亞於在可見的未來或許將完成的那個最後製作。

生產意義的人

這必然意味著工作倫理的轉變。在傳統的認識框架中，創造活動總是需要一些私密性（正如所有實驗需要有容錯的空間），對於過程公共性的企求往往顯得衝突，或甚至被認為僅僅是服膺於某種文化治理的工具理性。但在一個更寬廣視野中，我們有沒有可能設想一種有別於傳統的「藝術實驗」的文化生產，朝向一種「文化實驗」的工作倫理？

「文化實驗」需要一定程度非菁英的、異溫層的群眾作為生產意義的人。這裡並非重覆「人人都是藝術家」這種陳腔濫調，而是用創造性的手段，將他們從被動的、刻板的消費者／收受者角色中適度拉開，使他們能在某些時刻成為意義的創造者。在這個轉換中，藝術家扮演的可能是一種中介者的角色，調度著文化場域中的異溫層，讓進入計畫中的人們能產生意義。這需要某種形式的動員——不見得是直接的群眾參與，也包括在公共領域中創造事件、提出問題、啟動對話。

CREATORS 計畫試圖在重新連結文化生產過程的線索、意義與勞動者的同時，能進一步地讓原本那些文化產物的收受者，可以在過程中被拉進來，共同成為生產意義的人。藉此，我們或許可以讓創造性的文化生產，在不可避免成為一種建制之同時，仍可以保有開放的特質——其結構與輪廓，總是不斷在自我消解又被反覆釐定。CREATORS 計畫正是在這個動態的邊界上，思考一個創作研發支持系統的任務，並試著透過這方式，慢慢將一個由短暫的計畫週期定義下工作室，轉變成一個創造實驗的聚落。

關於這本書

本書以編年排序，收錄 2018 至 2020 年間在 CREATORS 計畫旗下的 44 項計畫及執行團隊（或個人）的介紹，及相關的展演紀錄，其間並穿插多位觀察員針對各項計畫的書寫文字節錄。書末則完整收錄藝評家王聖閎針對 CREATORS 計畫整體機制的重量級評論文章，以及一場從觀察員角度的 CREATORS 計畫回顧座談之紀錄。期望對於這三年歷程的捕捉，能呈現 CREATORS 計畫在持續滾動中的視野與精神，這也是 C-LAB 對於文化實驗的一份階段報告。

Preface: From Studios to the Community Generating Experiments

TEXT_YU Wei

Since the official launch of the Taiwan Contemporary Culture Lab (C-LAB) in 2018, the CREATORS Creation/Research Support Program (CREATORS Program) has been one of C-LAB's core development projects. CREATORS is a support system that integrates subsidy, residency, incubation, public activities, and the possibility of work commissions and exhibitions/performance for art and projects at their initial phases. With open calls every year, C-LAB supports those experimentative, innovative, and interdisciplinary projects using its own resources, and endorses their development and practice for six to nine months.

The project practitioners in CREATORS are, of course, recipients of subsidy contracts from C-LAB. However, in a more insightful interpretation of this relationship, they are also potent collaborators with C-LAB who produce contents and propose visions for this very base and who demonstrate the possible, dynamic definitions of cultural experimentation in their own pursuits.

Beginning as a program that only provided "residency" spaces, CREATORS program has been recalibrating its operations annually and is evolving toward a composite support mechanism to better meet the needs of contemporary art and cultural productions. In 2019, in addition to its original residency model, CREATORS Program added the Creation/Research Support category for creators who do not need space; this enabled the program to connect with other creative individuals rooted in different locations across Taiwan. Then, in 2020, the CREATORS Sound-Off: Up-and-Coming Creators Residency Program, which has a focus on Ambisonics sound creations, was also included in the program; this program promotes new talents of experimental sound and music with the final goal of producing collaborative productions with the Taiwan Sound Lab of C-LAB.

What is CREATORS experimenting with?

Compared with traditional, results-oriented subsidies that require a finished production, CREATORS Program instead values the significance of every action taken in the process of cultural productions and the potential publicness they bring about. In the past few years, CREATORS' development has also been concerned with how to establish a new discourse that shifts from being focused on "the result" to "the course of the result" and how to generate feedback that is constructive to the development of the program. This is why CREATORS Program was structured to use the observer system: every project was assigned a Consulting Observer — who was an expert in the relevant field and provided opportunities for consultation and discussion — to ensure the project would not be developed in isolation. Then, in 2019, the idea of an Reviewing Observer, who offers observation and critical feedback in writing, was introduced. The reviewers were then expanded into the Reviewing Observer Collective in 2020 with their writings became featured articles on *CLABO*, in which these "writings about the progress of the projects" also became an experiment of the existing art critic and report genres.

Ideas about the significance of the actions involved in progress, publicness, and interdisciplinary practices are greatly influenced by the exploration of, and reflection on, the concept and modus of "cultural production" in the practice of art, and the boundaries set by contemporary art are the key reference to the CREATORS program. Today, in the macro field of culture, contemporary art is still a relatively flexible subterrain that embraces imagination, as well as an ideal field for interdisciplinary experimentation. Through plastic, perceptive, and conceptual art production, some artists can apply their practices to society and individuals, while others can endeavor to ignite sparks between various disciplines and knowledge systems.

In CREATORS' projects, artists, being the role that is included in most of the proposals, are often far more than "creators" in the conventional sense; they are more like actors or agents of these extended, diverse cultural practices, and they engage in fieldwork with participants, host workshops, or share their research results through different discussion formats. There are also a considerable number of CREATORS recipients who are themselves intermediaries in the cultural production chain, such as editors, critics, curators, cultural actors, members of nonprofit organizations, or those within a self-organization.

While the subjects and compositions of the selected CREATORS projects from 2018 to 2020 were diversified, they also shared some common aspects. There were experiments on the language of art performances, including the body languages of contemporary circus, Nanguan, Vietnamese Ca trù, Nong Nao acrobatic performance, the rituals performed by the spirit medium tâng-ki, along with the exploration of new visual languages used in motion pictures, sheet music, speculative design, and the collage of found objects. There were also experiments related to the various applications of new media art, including point cloud scanning, virtual reality, real-time audio-visual production, emission detection, Wave Field Synthesis (WFS), genetic modification, algorithm, web crawler, Ambisonics, and human-machine creations with artificial intelligence.

The tendency of interdisciplinary cultural and historical research is also seen in projects of various themes, such as the diaspora experiences of Chinese mainlanders in post-war Taiwan, the history of stigmatization of homosexuality in Taiwan, volcanoes and geography/geology, Taiwan's sugar industry history, mining culture, songs during the Japanese period, as well as research that was rooted in media criticism, internet culture, and urban spaces. Meanwhile, some teams also chose to focus on self-organization and cultural actions, including creators' social platforms, settlement preservation, the reuse of vacant spaces, weed collection, fermentation and urban space, and urban gardening. Lastly, some of these projects were even developed out of the contemporary issues Taiwan is grappling with, such as information warfare, the Anti-Extradition Law Amendment Bill Movement in Hong Kong, and the COVID-19 pandemic. These projects all dared to step outside the ivory tower and deal with reality through differing routes.

When we think about studios

Thinking about the importance of the process of cultural production from the perspective of art necessitates a reflection on the history, physical practice, and essence of the "studio." A (artist) studio is an aesthetical, political, and social institution that, in its early development, may serve as a private venue for the artist's creative activities, a storage space for piles of materials and half-completed artworks, or as a hybrid space for exhibition, communication, and social gatherings. These definitions, however, were still based on the premise of art object productions.

As the conceptual artists of the 1960s and the 1970s radically redefined the idea of an "artwork," they also challenged traditional understandings of artists' studios. In the so-called "post-studio" era, some artists' studios were comparable to offices or factories with employed administrative assistants or working laborers, while others were like a node in a social network – a social locale that constantly engaged in various external relationships. Then, there were studios defined not by physical spaces but by laptops operated in a bedroom, café, or hotel lobby. Also, let us not forget how the position of artist-in-residency periodically provides artists with opportunities to travel, create, and research in a foreign land on a short-term basis; during this period, makeshift studios are created with the more social purpose of expanding interpersonal networks, and creative activities are fused with the globalized mobility experience.

Building on the development of the "studio" in the history of art production and the practice of contemporary art, we took the idea of an incubator for start-up businesses and applied more active components of project consultation and catalyst support to envision the possible formats the residency-based CREATORS Program could take. We decided the program should keep the modernist function of being a private space for creative activities but also serve as an open field for the expression and practice of creativity; it should be a quasi-social space with elements of publicness.

Through such a concept, we intended for the recipients of the CREATORS program to adopt various forms of public activities with greater society during their residencies. We hope that these activities will be more than merely promotional or educational events to meet the public expectations of "seeing the experiments"; we hope that they will be a profound and organic part of the creators' projects that allows for feedback on the projects' development derived from engagement with the public, and they will be a driving force motivating the creators to drive their projects forward or recalibrate them. To us, these public activities extending from the development of the projects are extraordinary, and their significance is certainly no less than the final productions that might come in the future.

Content producers

The demand for publicness in the process of cultural productions indicates the necessary changes of work ethics. According to the traditional cognitive framework, creative activities require a certain level of privacy – just as experiments require certain tolerance of failures – that is often in contrast to the demands of publicness of the process and considered as an instrumental rationality serving cultural governance. However, could we not broaden the horizon and devise another kind of cultural production that is detached from the idea of "art experimentation" and, instead, oriented toward the ethical approach of "cultural experimentation"?

"Cultural experimentation," to a certain degree, requires that the non-elite mass outside of the echo chamber be content producers. Here, the intent is not to endorse the "everyone is an artist" cliché but, rather, to pull the masses out of the passive consumer/receiver stereotype and recast them as content producers at certain moments. According to this approach, artists are cast in the role of the intermediary, regulating opposing opinions in the cultural field so that the people participating in the projects can also be represented. This calls for certain mobilization, which need not necessarily be in the form of direct public participation; this mobilization could also be achieved through creating events, raising questions, or facilitating conversations in the public realm.

CREATORS Program attempts to include those who were once receivers of cultural productions as collaborative content producers and to reconnect the clues, connotations, and laborers involved in the process of cultural production. Through this, we may find a way to allow creative cultural productions to retain their openness while inevitably becoming institutions—in which their structure and presence are always canceled and redefined within themselves. It is along these dynamic borders that CREATORS Program speculates about forging a support system for creators and their research and gradually transforming the project-oriented, periodically existing studio into a community generating experiments.

About the book

This book encompasses introductions to 44 projects endorsed by the CREATORS Program between 2018 and 2020. Introductions are also provided for the executive teams and individuals involved in the projects, and records of their exhibitions and performances are included, along with excerpts of the Observers' writings about the projects interspersed between chapters. In the last section of this book, influential art critic WANG Sheng-Hung's full article reviewing the CREATORS mechanism has been provided, along with a text transcription of the talk in which the CREATORS Program Observers reflected on this program. In this retrospection of these progress over the past three years, this book serves not only as a presentation of the prospects and spirits of CREATORS Program in its ongoing evolution, but also as a work-in-progress report of culture experimentation at C-LAB.

Archive or Alive ── 劉守曜獨舞數位典藏

Archive or Alive — Digital Archiving Development of a Solo Dance by LIU Shou-Yuo

本計畫「Archive or Alive ──劉守曜獨舞數位典藏研發」旨於應用當代數位科技，針對表演者的肢體動作進行非物質典藏技術研發與方法論建構。突破過往單視角影像紀錄，改採全視角的動能訊號捕捉，以環視鏡頭拍照，攝影表演者全身，如此產生的紀錄影像期能幫助觀者具空間感知地，以 3D 視角從不同視點觀看肢體動作。

The project "Archive or Alive – Digital Archiving Development of a Solo Dance by LIU Shou-Yuo" aims to apply the contemporary digital technology for the physical movements of performers in conducting the development of non-material archiving technology and the building of a methodology. It breaks through the conventional single-view videotaping and further captures the signs of kinetic energy in full view. With the holistic view of the performers, it hopes to assist the viewers to look at the three-dimensional physical movement from different perspectives with a sense of space.

CREATORS

在地實驗 ET@T

在地實驗由藝術家黃文浩創立於 1995 年，觀察與發展所有具有潛力的藝術形式，並探索因數位文化而產生的不明狀態。以客觀檢視、主動投入的雙重身分，持續延伸觸角到當代藝術各個可能的領域，積極發展成為集合理念與實作能力兼備的機構。

ET@T was founded by artist HUANG Wen-Hao in 1995. It observes and develops all the potential forms of artistic practices and explores the unidentified situation caused by digital culture. With the two-fold identity of an objective evaluator and an active participant, ET@T ventures into every possible discipline of contemporary arts and actively develop itself into an institution which obtains both concepts and implementing skills.

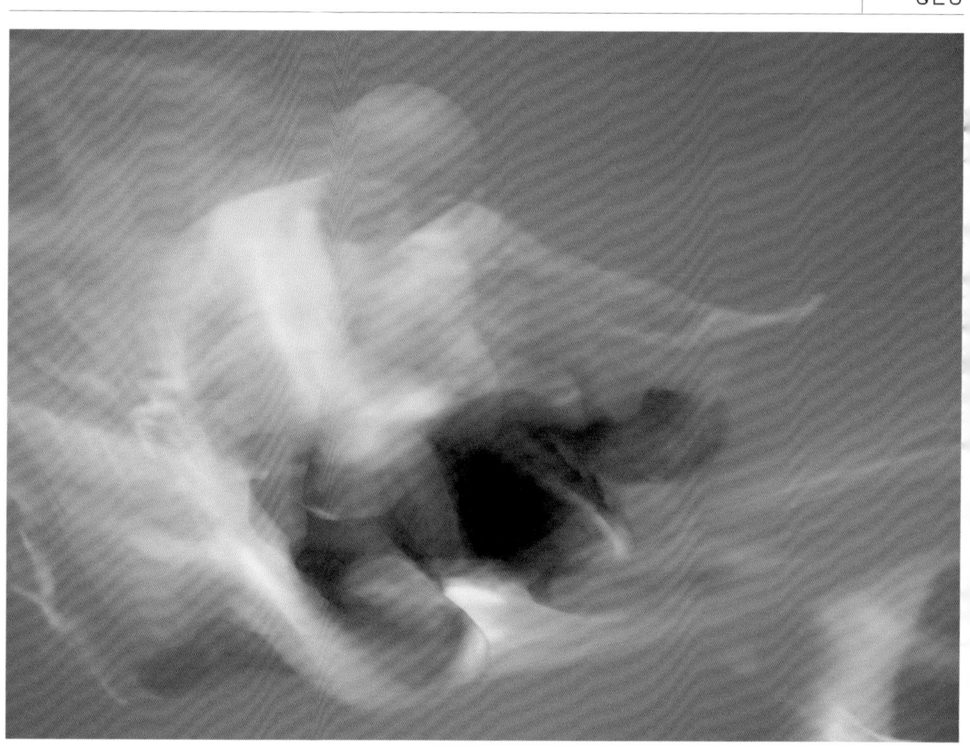

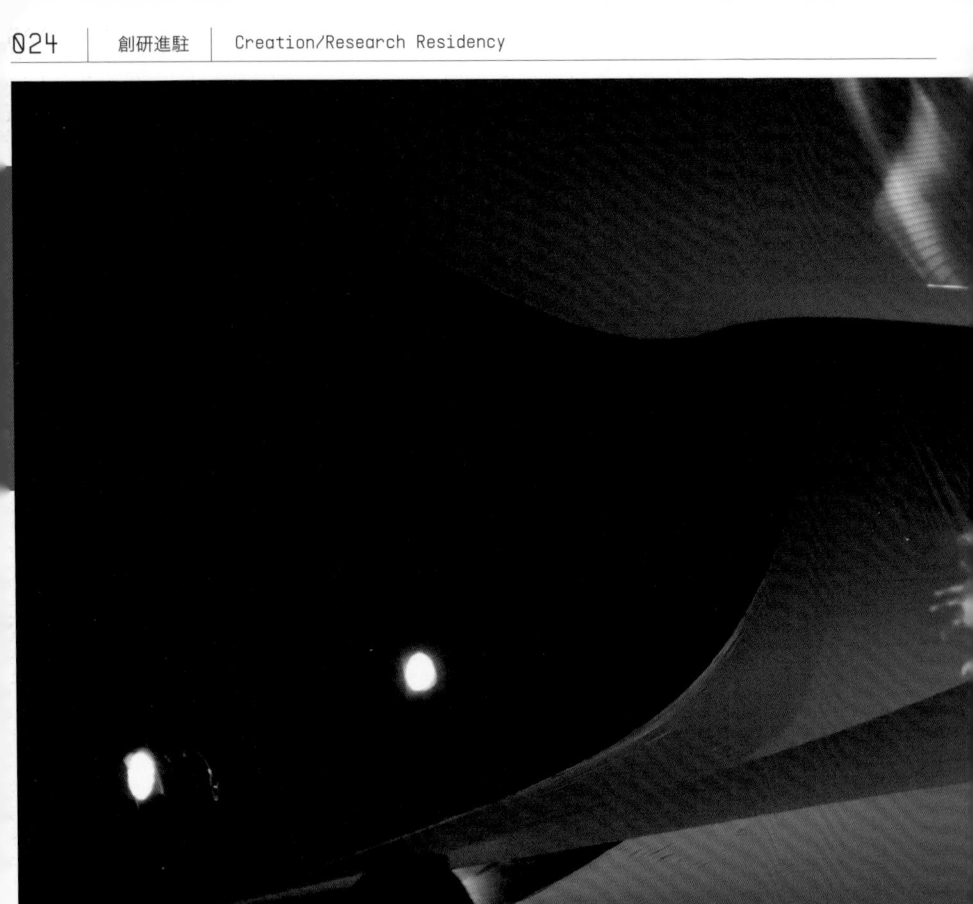

▌在地實驗，「Archive or Alive — 劉守曜獨舞數位典藏」，2018_

ET@T, *Archive or Alive-Digital Archiving Development of a Solo Dance by LIU Shou-Yuo*, 2018_

- 「Archive or Alive——劉守曜獨舞數位典藏研發計畫」是「概念美術館」針對臺灣的「行為」、「表演」、「身體」、「慾望」、「劇場」、「舞蹈」……等這些問題群的歷史考察的第一步，這個計畫的「記錄」同時在意的還有「研究者」在查找與思考資料上的「預期」。換句話說，相較於一般的搜尋引擎在資料來源與資料內容上的分散性，「概念美術館」相對而言針對的是專業的使用者（研究者），期待這些研究者因為擁有相對「知識密度更高」的資料，而省去每個研究者都必須重新整理資料的力氣，另一方面，也是藉由這些「知識密度更高的資料」之間的鄰近性，創造某些透過「比較」而來的知識再生產的契機。

  - 「概念美術館」並不以單一表演團體（或個人）的存續及發展為其主要考量，相反地，透過不同作品、表演者與導演編舞家之間的「比較」而生產出來的「知識」才是其主要的目的。

- 在「概念美術館」這樣一種媒介理論式的考慮下，「Archive or Alive——劉守曜獨舞數位典藏研發計畫」所提出的「紀錄性的重演」，既是舊的、也是新的。舊的部分指的是，這是一個立基於 2014 年於牯嶺街小劇場演出之《Shapde5.5》的「紀錄」。不管是在精神上，還是在空間裝置上，都希望盡量能夠貼近當時的設想，縱使因為時空環境條件的變化而不得不有些取捨，但是基於「研究」的需求，所以期待的並非「新作」。新的部分則是技術變遷及格式漂移的部分：VR 的影像技術到底要以怎麼樣的「紀錄」方式才能夠貼近原來的表演，而不會被認為是新作？

  - 團隊在這個計畫中暫時捨棄了使用「動作捕捉」的想法，改採全景式的紀錄模式，換言之，暫時不將「動作」看成相較於其他表演元素來說具有較大的資訊值，而是均值地看待所有的表演元素，也就是利用「空間」的包覆性效果，將整個作品「打包」在一個空間之內。（文｜王柏偉）

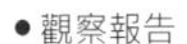
觀察報告

王柏偉〈在地實驗「Archive or Alive——劉守曜獨舞數位典藏研發計畫」觀察員手記〉，全文請見
https://mag.clab.org.tw/clabo-article/clab-creators-ett-archive-or-alive/

空氣結構 lab —— 在島上呼吸
Air Structure lab —Breathing on the Island

「在島上呼吸」分為三個實驗方向：一、複合材料的研發：透過不同充氣面料的研究實驗，使空氣密封於面料之中，達到快速充氣，快速洩氣，不漏氣，防水等效果。二、充氣面料的加工手法：透過對圖案與摺紙的研究，使面料產生結構性，藉由充氣，使 2D 的平面到 3D 的空間，洩氣後能快速收納。三、充氣面料與構建的結合：於充氣面料中置入小型構建與感應器，達到智能化的使用。

“Breathing on the Island” is an experiment with three-fold directions:

1. the research of composite materials: through the trials of different inflatable fabrics, the project aims to develop a fabric that can seal the air and achieve the features of quick inflation and deflation, being airtight and water-proof.

2. the processing method of inflatable fabric: through the studies of pattern and origami, the fabric will be folded and imposed with a specific structure. By inflating the structure, it will grow from two-dimensional plane to a three-dimensional balloon; furthermore, it can be easily stored after deflation.

3. the combination of inflatable fabric and components: putting small components and sensors in the inflatable material to reach the intelligence function.

CREATORS

空氣結構 lab
Air Structure lab

「空氣結構 lab」由徐業詞、余承龍組成，專門研究空氣結構，利用充氣、折疊的機制，創造出有意義的變形物體，使未來生活將變得更加方便。團隊致力於研究空氣結構與其應用發展的可能性。

“Air Structure lab” was founded by YU Cheng-Lung and HSU Yeh-Tzu. It specializes in air structure research and makes use of the mechanism of inflation and folding to create meaningful deformed objects for making future life more convenient. The team devotes to studying the structure of the air and the possibility of its application and development.

▌空氣結構 lab，「空氣結構 lab — 在島上呼吸」，2018_

Air Structure lab, *Air Structure Lab-Breathing on the Island*, 2018_

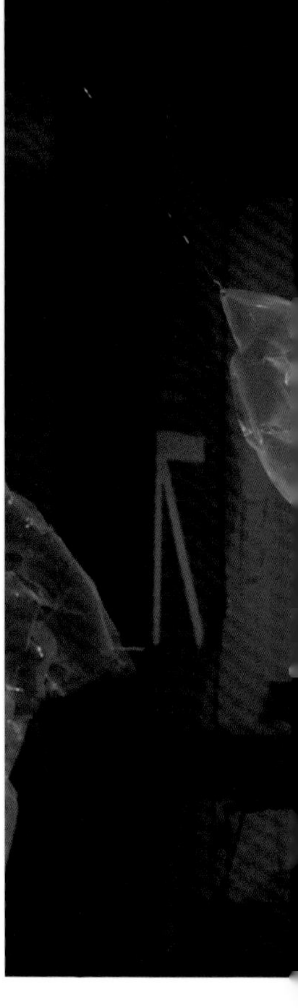

> 如果說建築是與自然環境的互動和對話，那麼空氣作為環境的一部分，要如何才能被人所感知？除了起風時刻，空氣是否必然遭到忽略？如果人們在空間中開始被空氣擠壓，例如以感應器讓充氣結構伸縮，擾動人們的行走動線，影響他們身體的感受，會創造出一種什麼樣新型態的環境現實？身處其中，人們又將如何以不同的方式存在著？

> 建築訓練更注重平面繪圖和結構創造，而為了創造動力空氣柱，業詞必須從頭學習電機基礎和電腦程式語言；而承龍則需要系統性地研究空氣結構的塑形規則。在工作室中，兩人乍看之下更像科學家，把主要的心思放在技術的掌控和改良。但再思考，如果油畫家需要實驗礦物顏料的調配、色度，以及反覆做人體素描練習，那麼以空氣為顏料的他們，反覆研究機械能耐，甚至創造實驗組和對照組系統以比較不同空氣柱的表現，不也是某種「素描練習」嗎？而素描練習的最終目標，他們說，是「為枯燥的居住環境提供新的想像，給予人們一種刺激，充實城市的面貌。」他們問：以後的房子可不可以帶著走？可不可以三秒成形。空氣提供了兩人一個想像力的出口。

> 空氣結構中存在許多其他材料沒有的張力和驚喜。空氣結構的體積與重量不必然有正向關係，看起來巨大的物件，實際上輕如鴻毛，可以瞬間移動。而原本平面的塑膠材料，在充氣之後能快速轉變為 3D 物件，洩氣之後又再次回到 2D 狀態，其存在具有維度上的模糊特性。再來，他們的空氣結構尚未存在既有演算公式，因此只能直接用實體進行測試。由於無法意料結果，每一次的測試都帶來驚喜。看似未來而前瞻的素材，其實有著非常不數位科技的傳統實驗過程。（文｜王莛頎）

● 觀察報告

王莛頎〈不存在的存在：「空氣結構 Lab」駐村計畫〉，全文請見
https://mag.clab.org.tw/clabo-article/clab-creators-air-structure-lab/

悲傷ㄟ曼波 ——臺灣弄鐃文化的馬戲與錄像創作

Melancholy Mambo —A Circus Video Work of Taiwan's Cymbal Playing Culture

本計畫目標在於田調採集與研究馬戲身體的同時，更以「dance video」（舞蹈影像）的錄像創作概念，重新探索與嘗試「circus video」（馬戲影像）的創作形式，結合弄鐃科儀表演與馬戲雜技，並和弄鐃團體的合作共創，未來將以劇場創作演出和錄像創作與展覽等形式，呈現出臺灣當代馬戲與弄鐃傳統表演跨域實驗的新風貌。

The goal of this project is to re-explore and experiment a new practice of "circus video" with the concept of "dance video" while doing field-work and conducting research on body movement of Circus performance. In cooperation with cymbal playing troupe, Thunar Circus will further combine cymbal playing rite with circus performance, thus presents a cross-disciplinary experimental performance of Taiwan's contemporary circus and traditional cymbal performing in the form of theater art, video, and exhibition.

CREATORS

圓劇團 Thunar Circus

圓劇團是一根生於臺灣的國際當代馬戲創作團體，專注於藝術表現及詩性的演出。希冀發揚當代馬戲的精神，並拓展表演者的創造力，在當代馬戲的舞台上，持續提供真實具藝術深度的表演。

Thunar Circus is an international contemporary circus creative group born in Taiwan. It focuses on artistic and poetic performance and aims to promote the spirit of contemporary circus. It also spares no effort to promote performers' creativities, so as to contribute to artistic depth on the stage of contemporary circus.

圓劇團，「悲傷ㄟ曼波 — 臺灣弄鐃文化的馬戲與錄像創作」，2018_

Thunar Circus, *Melancholy Mambo – A Circus Video Work of Taiwan's Cymbal Playing Culture*, 2018_

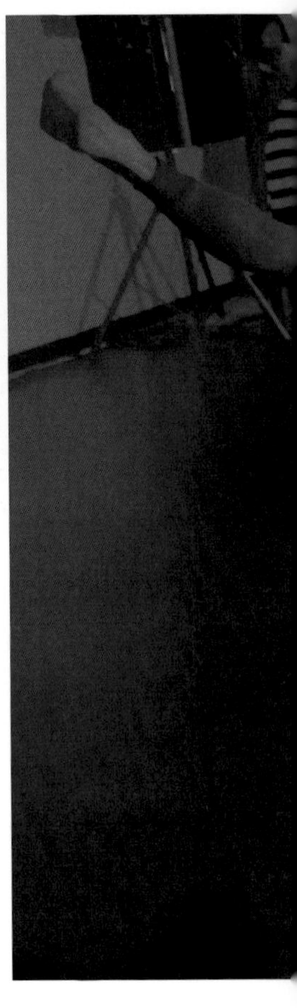

> 喪家通常在「三七」(作旬) 請來弄鐃，相較於其他儀式，弄鐃不具有超渡功能，只調節悲傷的氣氛。師傅的服裝，從色調到扮著皆十分樸素，黑衣服白褲子，卻在色彩鮮艷的喪葬場合放很跳的音樂，舞很鬧的鑼鈸，反差甚大，屬於喪葬科儀中的特技表演類陣頭，像牽亡歌陣、已經佚失的三藏取經皆屬此類；若為兩人同行，則一人表演一人音控加旁白，表演讓人驚呼連連，吉祥話讓人慰藉滿滿，師傅亦會感覺、觀察現場喪家的情緒，即興地調整自己表演、說話的尺度。

> 弄鐃既然是在開放的空地或廣場上表演，圍觀的群眾也就並非購票進劇場，篩選過的觀眾，而是混雜各色背景的人，這讓人想到即興與場域的關係。當問及學習弄鐃對於當代表演的啟發，正宗提到其一是「隨機性」，回顧學院雜技教育及創團至今的歷程，他認為必須先把「街頭」和「即興」切開，兩者不等同，練雜技的學生會想辦法去街頭表演，但如果只是偶一為之就很跳脫既有習慣，累積經驗、摸索方法。他觀察到，現在街頭藝人有很多新的發展，像「街頭作品實驗室」就在嘗試街頭表演的可能，不是透過大量且刻意的互動，而是藉由技術的純熟及趣味吸引觀眾。對他來說，取徑民間文化發展當代馬戲、劇場創作，主要的動力還是從自己出發，因而回溯童年經驗找回弄鐃。

> 「實驗」往往是新鮮的誘惑，帶有一種「過去的什麼都不要」的叛逆性格，但戒嚴背後的冷戰體制卻是一口氣把「藝術」與「民眾」切斷，因此，民眾和藝術其實都是 1980 年代的產物，宛如新生。於是，不斷遺忘、拒絕過去的「趨新」成了我們文藝發展的標記，發展即斷裂，我們和我們追逐的夢孰長。一個又一個詞語入境，我們卻總是來不及記憶，於是只好把「在地」漆上失憶而封閉的島嶼保護色；陳傳興的延遲現代性及論述沙漠化之說、王墨林的身體論等，與其說是在努力記憶，不如說是在探勘社會的集體失憶有多深。(文 | 吳思鋒)

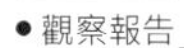

● 觀察報告

吳思鋒〈跟悲傷的人一起跳舞：火，以及眾的平面〉，全文請見
https://mag.clab.org.tw/clabo-article/clab-creators-melancholic-mambo/

回聲像—動態影像論壇與音像創作實驗計畫
Audiovisual Return — Moving Image Forum and Creative Audiovisual Experimentation Project

此項進駐計畫「回聲像—動態影像論壇與音像創作實驗計畫」，以回顧過去臺灣動態影像的歷史經驗為初衷，討論數位語境下的動態影像與音像表演創作之現狀、為未來提出值得參照的臺灣觀點。針對當下臺灣動態影像的定義與形式、動態影像與數位故事創作、音像創作實踐動態影像的感官探索，以及動態影像的未來等主題進行論壇。進駐期間亦組織一組跨域創作團隊，以「即時電影」音像創作形式作為實驗手法，以實際參與論壇、回應對話為主旨，委託創作一件結合動態影像研究、實驗形式與數位藝術的展演作品。

This project starts from reviewing the historical experience of Taiwan's historical moving images and tries to tackle the current situation of audiovisual performances and practices in digital context, proposing a local perspective for the future. Forums will be organized according to issues such as current definitions and forms of Taiwan's moving images, practices of moving images and digital stories-telling, the exploration of senses in audiovisual works of art, and the prediction of moving images. The project builds a cross-disciplinary team, which takes "real-time cinema" audiovisual art-making as a experimental method to participate in the forums and respond to the issues aforementioned. The team further creates a work which combines studies and experiments of moving images with digital art practices.

CREATORS

ART SHELTER

ART SHELTER是一藝術家與策展人獨立籌組的當代媒體影音平台，長期致力於動態影像與實驗音像類型藝術之國內推動與海外交流，加強理解新時代語彙，探索跨越科技與藝術邊界的各種可能，並以獨立方式建立起國內以音像為主題的「Osmosis滲透媒體影音藝術節」。

Comprised of a group of artists and curators, ART SHELTER is a platform of contemporary audiovisual media. It has long been committed to the promotion of moving images and experimental audiovisual art in Taiwan and abroad. It also strengthens its understanding of languages in new generation and explores every possibility that crosses borders of technology and art. It also founded independently OSMOSIS Audiovisual Media Festival.

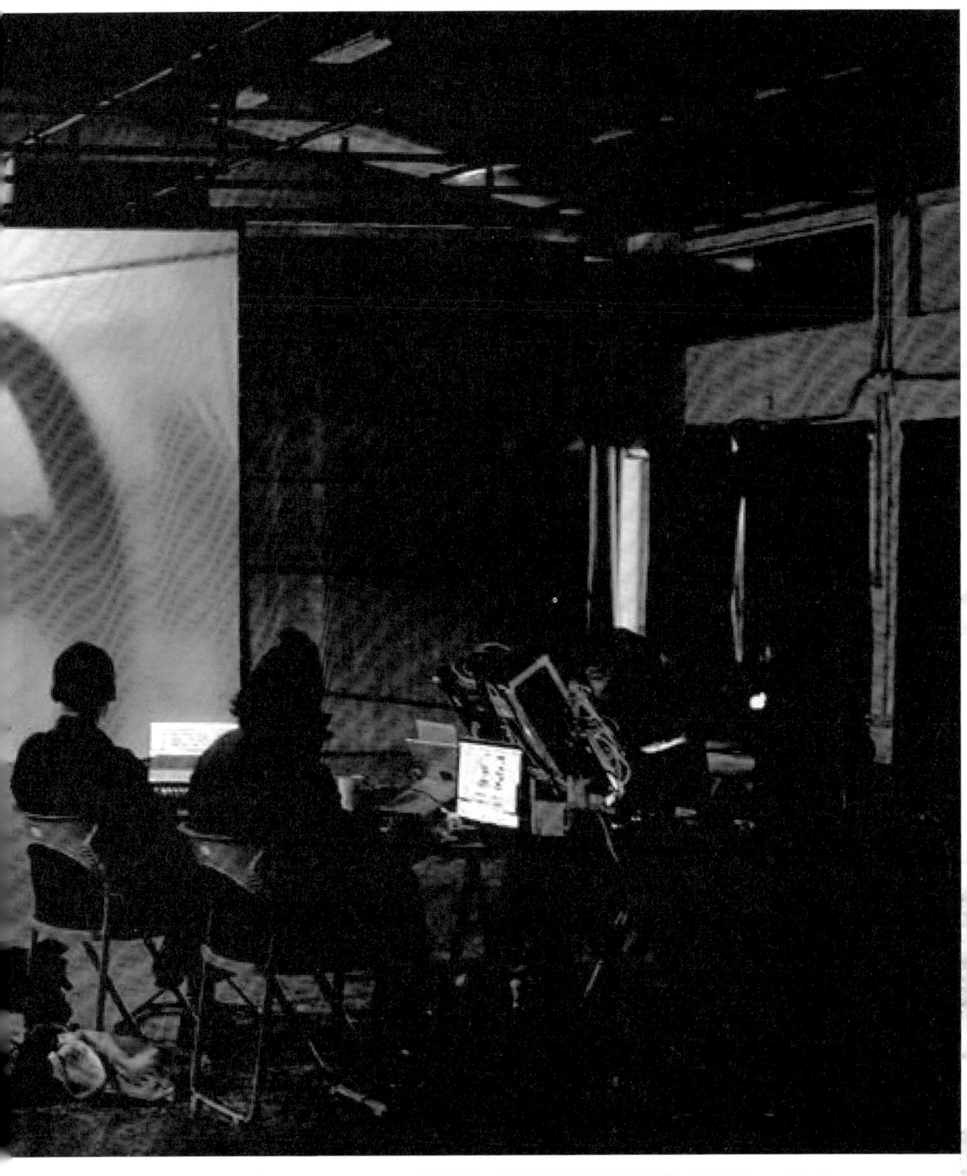

ART SHELTER，「回聲像—動態影像論壇與音像創作實驗計畫」，2018_

ART SHELTER, *Audiovisual Return - Moving Image Forum and Creative Audiovisual Experimentation Project*, 2018_

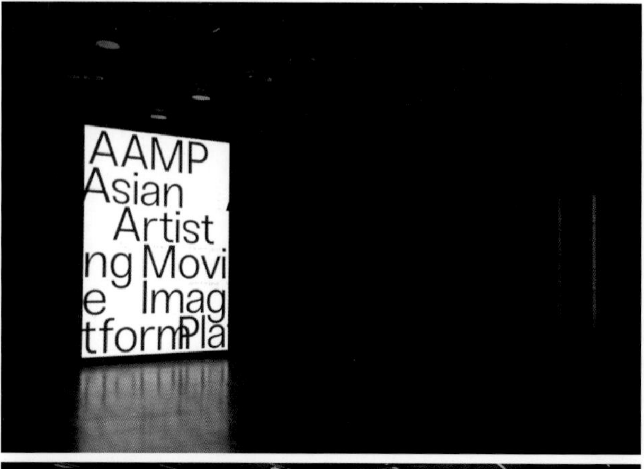

> 「回聲像」的研究創作計畫是以三場主題式的座談討論為研究框架，終以音像新作發表為架構的論壇與創作實驗計畫。「實驗」這字眼其實很有意思，它是個雙面刃，固然有新創研發、令人期待的產出，同時也隱含著可能會失敗、或者結果不如預期的風險。在實驗場域進行著實驗計畫，我們想突破什麼沉痾？想走去哪裡？能承擔得起哪種程度的失敗？能承擔得起走到哪個階段的新創實驗？由此來看回聲像三場論壇的規劃：第一場論壇討論的是根本性的問題「什麼是動態影像？」，循序到第二場探討「音像形式如何實驗作為感官研究方法？」，終以探索動態影像的未來可能為現階段研究句點。三場座談的規畫從本質性的提問探討出發，到現下正在發生的場景，終至探問未來。該如何在每一個往前進的時代發展節點上展開所有可能性，並折返回到起點，回應最根本的問題——影像是什麼？什麼是動態影像？（文｜蔡家榛）

● 觀察報告

蔡家榛〈實驗．影像．對話：回聲像計畫觀察報告〉，全文請見
https://mag.clab.org.tw/clabo-article/clab-creators-art-shelter/

沉浸式逝去影像考古計畫
The Archaeological Project of Immersive Elapsing Images

此項計畫使用 3D 掃描和沉浸式數位載具，嘗試記錄個人故事和當地歷史，將 1：1 的空間模型置入在 3D 繪圖軟體中重新編排、建構，以便觀眾體驗虛擬的真實場域。本計畫希望建立更具相容性的規劃製作，利用 3D 體驗技術直接連接社群和開發人員進行對話·用數位紀錄保存即將成為歷史文件的空間和故事。

This project aims to document personal stories and local history with 3D scanning technique and immersive digital devices. This project processes the real-size model of the space with 3D modeling software for reconstruction to allow the audience to experience the virtual space. In such a way, this project aims to establish a more compatible production, connecting the community and the developers for direct conversation, and preserving the archive-to-be spaces and stories utilizing digital tools.

CREATORS

黃偉軒 HUANG Wei-Hsuan

黃偉軒作品包含多種形式，如劇場影像、聲響、聲音影像。作品曾受邀至美國科羅拉多州林肯中心、國立臺灣美術館、伊日藝術臺北空間及南海藝廊等參展。曾參與失聲祭、混種現場聲音影像演出，並擔任河床劇團、狠劇場、兩廳院製作等影像設計與執行。

HUANG Wei-Hsuan's artistic practices vary in forms, such as theater visual design, sounds, and audiovisual. He was invited to present his works in the Lincoln Center in Colorado, USA, National Taiwan Museum of Fine Arts, YIRI ARTS Taipei Space, and Nanhai Gallery. For live audiovisual performance, he has participated in Lacking Sound Festival and On Site. Also, he served as visual designer for Riverbed Theater, Very Theater, National Theater & Concert Hall, etc.

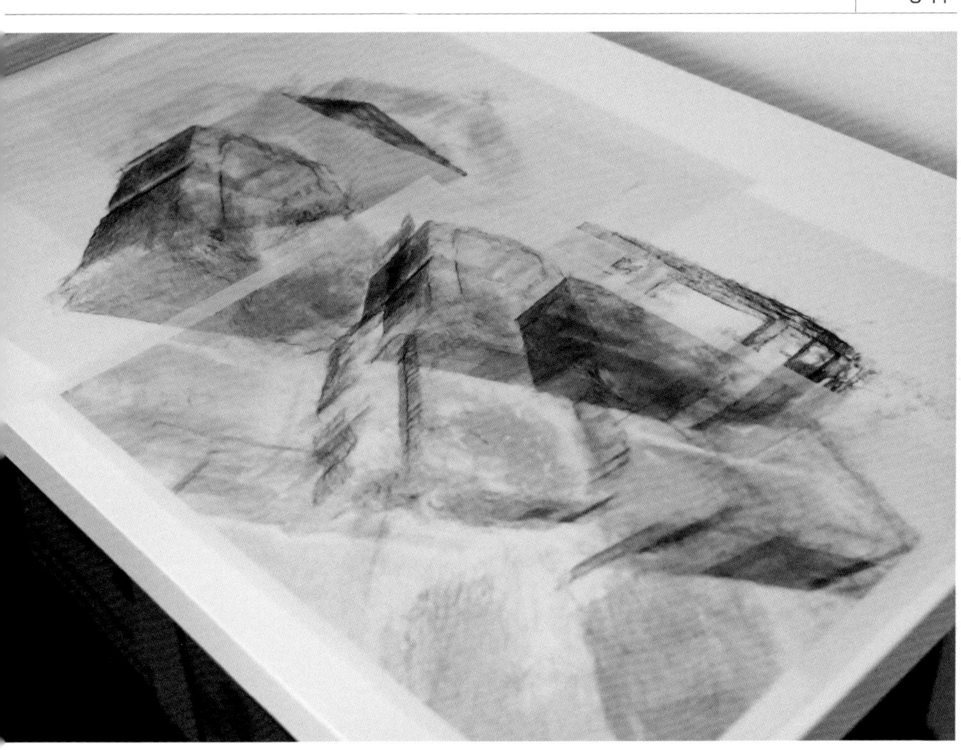

黃偉軒，「沉浸式逝去影像考古計畫」，2018_

HUANG Wei-Hsuan, *The Archaeological Project of Immersive Elapsing Images*, 2018_

> 「沉浸」所指的是消除觀看 / 演出的範圍邊界，令觀眾得以直接沉浸於作品之中的方法（反之，即作品直接承認觀眾的存在，在某種程度上觀眾也被放入其中）。它可能藉由特定位置讓觀眾與演出者交談、與周圍環境的互動甚至直接參與敘事等方式達成，主要立基於觀眾的主動性參與，而真實空間、感覺、運動與時間四個部分，則是構成其「沉浸」感的要素。而在黃偉軒的計畫裡，「沉浸」（immersive）一詞則被擴充為空間構成的概念，它既可與其他創作形式結合，亦可單獨形成可被閱讀的時間敘事。他在此首先提出的問題意識是：「在數位時代中，保存空間記憶的可能性為何？」約自 2013 年起，他開始嘗試將影像以電腦運算轉換為 3D 空間資訊的方式創作，藉此反覆回應他所思考「記憶與空間」兩者的連動關係。

> 黃偉軒由 3D 點雲輸出的空間，保留了現場光影甚至地面灰塵和各種殘留物的起伏，而這種構成方式，也沒有如一般空間建模時所具有的假設性全知視角，黃偉軒的空間模型不指向對建築物的全然透視，而是與記錄者的身體運動、目光所及範圍有關。

> 「沉浸式逝去影像考古計畫」應是目前對空總空間進行數位測繪最密集、累積空間數量最多的創作計畫。在此計畫的前期階段，黃偉軒透過攝影測量法（Photogrammetry）將攝影記錄後的空間場景後製為 3D 立體點雲，再藉由沉浸式數位載具提供觀眾一虛擬的真實場景。在進駐過程中，除了個人的創作內容外，黃偉軒也反覆強調該項空間掃描方法與其資料庫的雲端開放概念，在幾次的工作坊教學中亦不藏私地提供技術分享，希望藉由一般民眾也能上手的基礎工具（手機、一般等級的相機與開放下載的線上軟體）、簡化技術門檻，以達到可被輕易擷取、且具擴散性地傳播模式，鼓勵以數位紀錄、保存即將成為歷史文件的空間和故事。（文｜林怡秀）

● 觀察報告

林怡秀〈擴張記憶的空間：再訪記憶與目光所及之處〉，全文請見
https://mag.clab.org.tw/clabo-article/immersive-past-image-archeology-project/

譯譜者：在譜間轉譯的研究創作計畫
Transnotators: A Research-Oriented Art Project of Notation

轉譯，必定充滿衝突，轉譯，也經常是被認為是不可能的，然而，藉著不斷的轉譯，並在轉譯時不察譜所預設的認識系譜。「轉譯」事實上能成為——也必須是種「創作」。

「譯譜者：在譜間轉譯的研究創作計畫」為一研究導向的創作計畫，即是試圖考察各類的「譜」及其所預設的認識系譜，藉著轉譯，去擾動既有的認識，從而觸發創作。「譯譜者」邀請鍾玉鳳、黃思農等音樂家，一起作為譜的轉譯者，在各自的樂器傳統（例如琵琶、胡琴、手風琴），與各類譜背後的認識系譜間進出，以創作新「譜」。在譜的轉譯間創作音樂，也將在研究、分析、轉譯與發表的各階段，邀請其他研究者、音樂家或舞蹈家一起參與。

Translation, which must be full of conflicts, is often considered impossible. However, by means of constant translation, one can comprehend the genealogy without presupposition during the translating process. In fact, "translation" can be – and should be a "practice of art."

This project is research-oriented. It investigates various "music score" and the presupposed comprehension of the genealogy of musical genres. Through transcribing, the artist tries to disturb existing understanding and further stimulates creative impulses. Musicians such as CHUNG Yu-Feng and Snow HUANG are invited to join this project. They work as transcribers and create new "music scores" with the instruments they play (such as Pipa, Huqin, and accordion) in the context of the genealogy of musical genres. Besides music-making, the project include other researchers, musicians, or dancers at each phase of research, analysis, transcription, and presentation.

CREATORS

謝杰廷 HSIEH Chieh-Ting

舞蹈音樂研究者、音樂家、藝術家。曾與音樂家大竹研、早川徹合作，其創作曾於臺北市立美術館、柏林 Galerie im Turm 展出。近年於德國從事音樂舞蹈研究，研究興趣涉及從現象學與文化技術觀點探察音樂與舞蹈的身體感、力動、記譜等，其書寫散見《劇場閱讀》、《表演藝術雜誌》等。論文曾於國際音樂學會（IMS）、國際傳統音樂學會（ICTM）等學術研討會發表。

HSIEH Chieh-Ting is a dance music researcher, musician, and artist. He collaborated with Ken Ohtake and Toru Hayakawa. His works were exhibited in Taipei Fine Arts Museum and Galerie im Turm in Berlin. In recent years he engages in dance music studies in Germany. His research interests are in bodily perception, energy, and notation of dance from the perspective of phenomenology and cultural technique. His writings can be found in *Performing Arts Forum*, *Performing Arts Review*, etc. He published academic essays at conferences held by organizations such as the International Musicological (IMS) Society and the International Council for Traditional Music (ICTM).

謝杰廷，「譯譜者：在譜間轉譯的研究創作計畫」，2018_

HSIEH Chieh-Ting, *Transnotators: A Research-Oriented Art Project of Notation*, 2018_

> 杰廷說：「當我們看到一張譜時，我們必須要去解它，它總是代表著一套系統、一套認識的方法。包含了一套關於時間、空間的認識。」

> 「譯譜者」是個無前例，試圖結合研究、轉譯研究素材為創作素材，並透過實作產出與其假設（hypothesis）對話的展演。這是一個具企圖心的實驗，從不同展演形式的記譜、記號及其再現著手，且擴張「譜」的定義至暗碼、宗教圖、方法導引，創建多重的思考路徑，將「譜」帶離單一的、封閉的釋譯系統。對於謝杰廷來說，「譜」在舞蹈、音樂兩個表演藝術領域中，都是不可缺的聯繫媒介。但他對於譜的興趣，卻不止於西方音樂與舞蹈的記譜，而拓至世界不同文化體系下的表演藝術記譜，甚而棋譜、天象圖、宗教圖、食譜，擴張了「譜」的文本（text）與脈絡（context）的範圍。

> 杰廷以「生、死」的概念帶入表演與記譜的關係。現代西方強調表演的「在場性」，將表演視為會消逝的事物，表演者與觀眾的「在場」，是為「現場」（live performance）的因素與經驗。若以此方式理解「演出」（performance）的「生」（live），譜可以說是「死」的。而德希達也認為「書寫」或「記寫」這件事，總是關於死亡、身體的。

> 由於參與「譯譜者」的三位藝術家的領域與創作路徑不同，為了這個計畫，他們除了有各自領域的書籍須涉獵，也列了相當數量的共同閱讀與討論的書單。除了計畫中安排好的進程，如「工作坊」、「開放的譯譜室」以及總結這個計畫的講座式展演外，他們三人也密集地以譯譜室為閱讀、討論、知識與思想交流的空間。這個過程，除了為了探造一條抵達某種未知的、另外的「譜」的可能，也是「譯譜者」計畫的實踐方法，而這樣的實踐方法包含了跨領域的田野與研究，除了見證這個計畫的參與者對於「譜」的反覆論證外，也見證了譯譜室作為一個具有實驗能量的空間，在短時間匯入了各方各家的思想與提問。（文｜鍾適芳）

● 觀察報告

鍾適芳〈「譯譜者」計畫的觀察記錄〉，全文請見
https://mag.clab.org.tw/clabo-article/clab-creators-transnotators/

Audiovisualizer in Da House.

此計畫「Audiovisualizer in da house.」於進駐期間內，固定時間在線上直播數位音像節目，聲音類型不限於電子曲風，在不影響節目的前提下，開放讓線下的閱聽者進入工作室參觀、觀察表演者與工作者。同時讓原本是以文字型態紀錄的創作者訪談，能夠走入鏡頭前，用影音呈現的方式紀錄與燧人氏編輯對談分享的過程。進駐後期則舉辦了一場以數位音像為軸，於數位藝術家設計的舞台上，展演進駐期間累積能量的 audiovisual 派對展演活動。

This project broadcasts an online live audiovisual show, which presents music genres not limited to electronic music. With the premise that visitors don't disturb the show, the audience is allowed to visit the studio and see how the show is made. Furthermore, artists are invited to share their experiences with Zuirens and interviews are videotaped into digital archives. In the latter stage of this project, Zuirens organizes an audiovisual performance party that takes digital audiovisual art as its core concepts, cooperating with digital artists who will be responsible for the stage design, and closing up the project with accumulated experiences and energy.

CREATORS

燧人氏 Zuirens

「燧人氏」創辦之初（2006 年）為一圍繞 VJ 文化為主軸的線上電子刊物。2008 年前後連結本地優秀的電子音樂創作者、樂隊、視覺及數位音像藝術家一同共演，陸續舉辦以 VJ 視覺、live audiovisual 演出的派對展演。同時納入電子音樂與數位聲響領域，多面地介紹 audiovisual 相關的藝術、作品及工具。2017 年「燧人氏」展開直播平台計畫，運用自行開發的播放器製播「Living Moment」的 audiovisual 直播節目，由視覺藝術家將 DJ 演出的畫面即時擷取進視覺系統中，加入與聲音鏈結的參數濾鏡。

When Zuirens was founded in 2006, it used to be an online magazine focusing on VJ culture. Around 2008, it organized VJ and live audiovisual performance parties gathering excellent local electronic musicians, bands, visual artists, and digital audiovisual artists. Meanwhile, it included the fields of electronic and digital sound, broadly introducing the practices, works, and tool of audiovisual art. In 2017, Zuirens launched a livestreaming project with a self-developed player to make the audiovisual live show, Living Moment. Visual artists instantly captured the DJ performance and processed with a visual system, applying a parametric filter of sound.

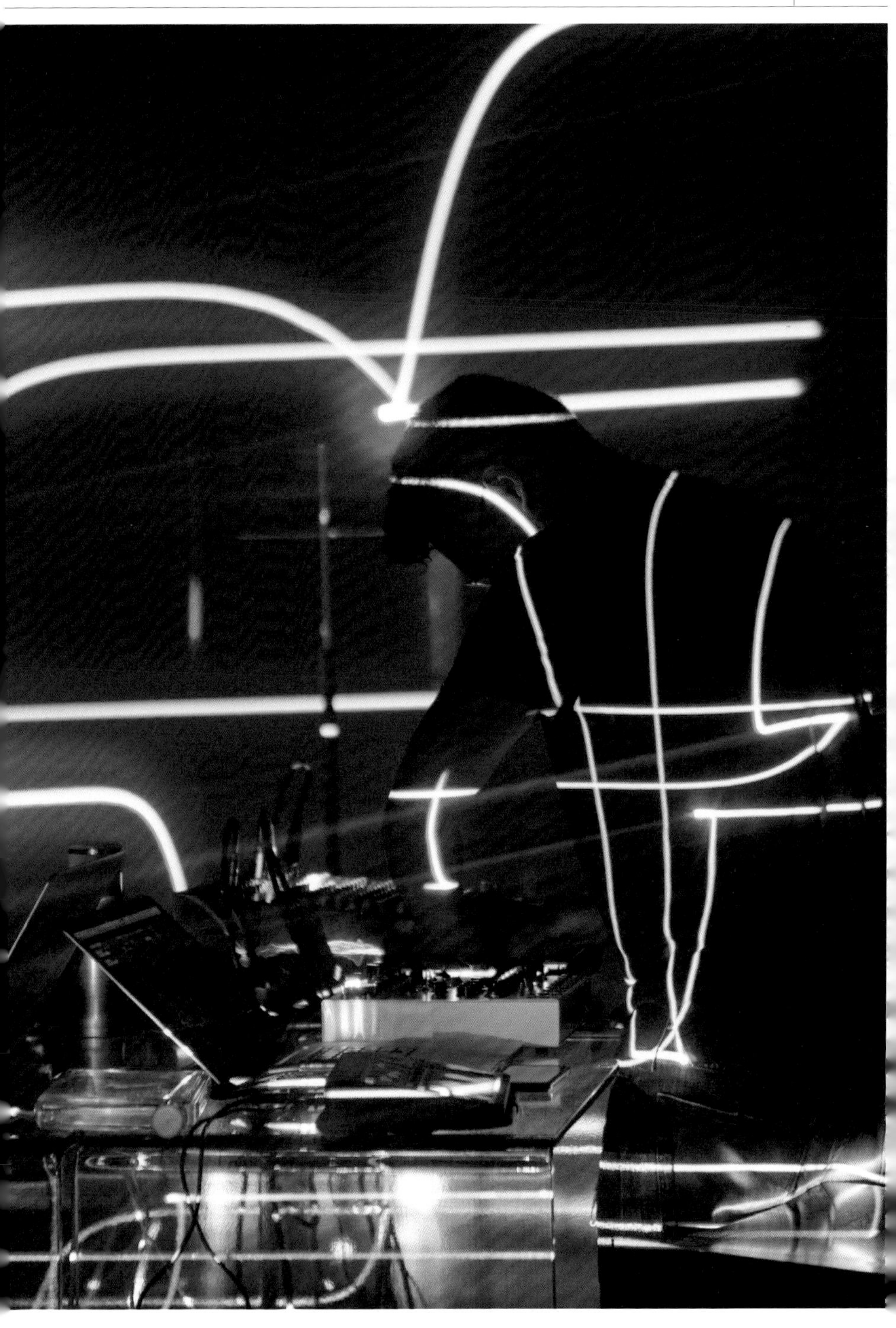

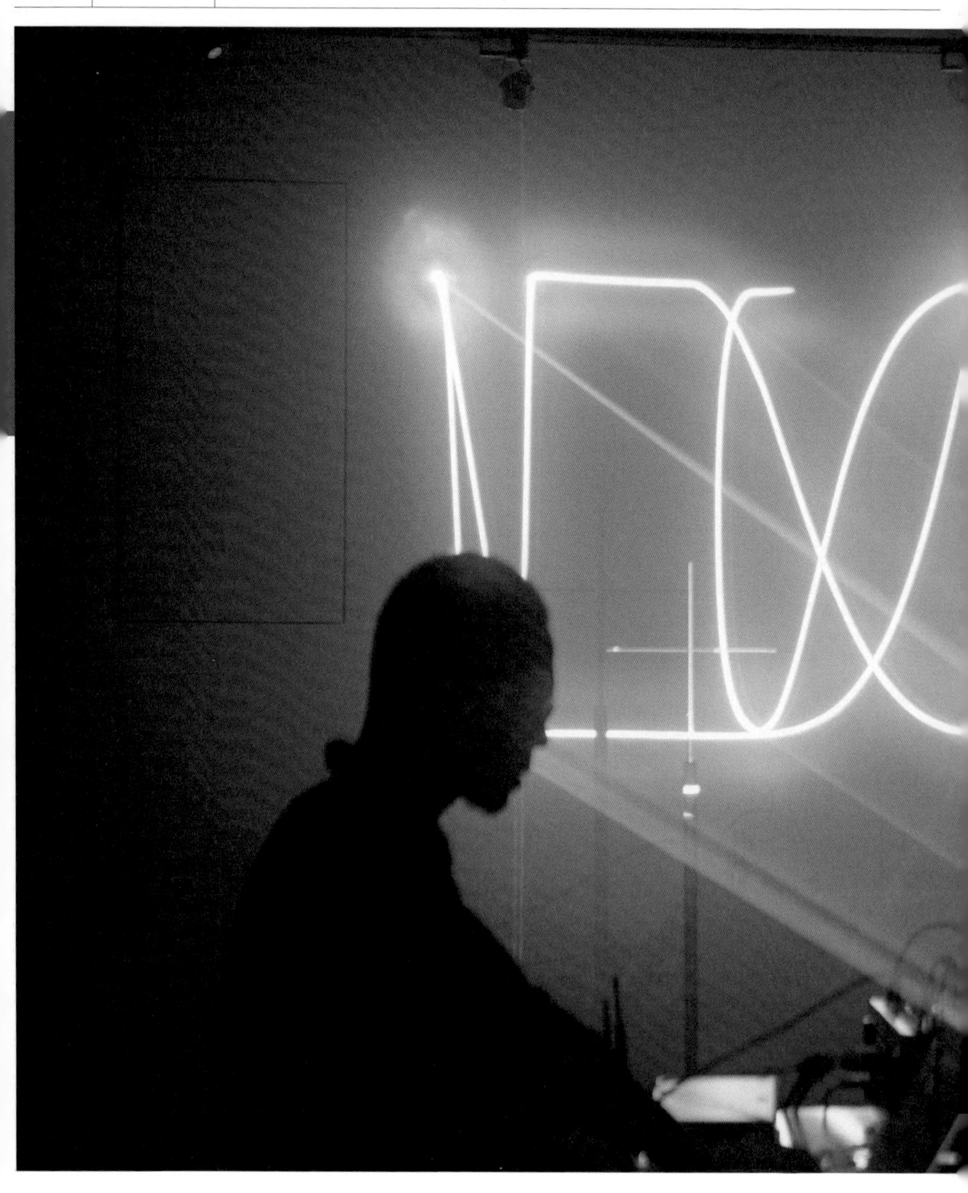

▌燧人氏，「Audiovisualizer in Da House.」，2018_
Zuirens, *Audiovisualizer in da house.*, 2018_

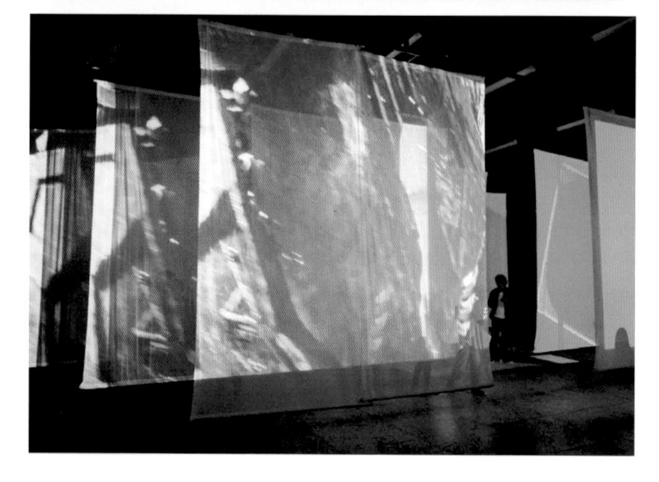

燧人氏，「Audiovisualizer in da house.」計畫網頁
https://clab.org.tw/creators/2018-r7/

「Lands」殘片文化實驗計畫
"Lands" Cultural Experiment Project of Fragmented Maps

「Lands」殘片文化實驗計畫，是以地圖連結作為依據，視殘片為理解當地變遷的重要考證，探討歷史與空間的演變歷程。這次計畫採用1957年〈臺北市街圖〉作為樣本，以空總為地圖中心點，東西南北方向進行文化殘片田野調查，每條路線由自然素材到科技產物都是取樣材料。透過殘片的收集進而發展成為創作素材，再拼貼構築成一個結合雕塑、繪畫、殘片文化詩的視覺裝置場域。

This project is based on the links of maps. Fragmented maps are important testimonies to comprehend changes and development of certain places. They can be used in exploring the developing process of history and space. This project takes "Taipei Street Map" (1957) as a case study and conducts fieldworks in cultural fragments radiating out from the location of the former Air Force Command Headquarters, which serves as the center on the map. The samples covers widely from natural elements to technology products found on each route; while the collection of fragmented maps will be further developed into materials for this project. Finally, a visual installation site combining sculptures, paintings, and fragmented cultural poems will be constructed.

CREATORS

劉時棟 LIU Shih-Tung

劉時棟1970年生於臺灣苗栗，畢業於臺北藝術大學美術系創作碩士班第一屆，長期從事視覺藝術創作。擅於透過拼貼、解離、並置、融混等手法，重組生活物件，創作蘊含豐富性與詩意。曾舉辦多次國內外個展及參與多次國際展覽，更榮獲多項國際藝術獎助與駐村創作邀約，深受國內外藝壇關注。

LIU Shih-Tung, born in Miaoli in 1970, holds a BA from Department of Fine Arts, Taipei National University of the Arts. He has long been working on visual art. He is skilled at reassembling everyday objects with methods of collage, dismantling, juxtaposition, and fusion which make his work rich and poetic. He held several solo shows and participated in many international exhibitions; also, he was awarded many international art fundings and received invitations for artist-in-residence while both international and domestic art fields.

特選檳榔

GOLDEN

▌劉時棟，「『Lands』殘片文化實驗計畫」，2018_

LIU Shih-Tung, *“Lands” Cultural Experiment Project of Fragmented Maps*, 2018_

> 長期以來，劉時棟的創作一直保持著某種細碎化資訊組合構成的特質。這些細碎化的資訊浮屑，讓藝術家得以拆解又重構了「歷史」和「記憶」的敘事架構與實相。從過往藝術家將這些細碎化的視覺及物件資訊，統合在深刻的傳統美學意識及繪畫結構中，到靜態人體 model 的拼貼，乃至於《殘片人》的誕生，其創作始終是「切身」所遇、所感的積累與構形，每一片浮屑都承載著劉時棟的生活及身體運動；與此同時，也承載著一個空間定點曾經的故事與事件。於是層層疊疊的拼貼成為了某種神祕的、隱喻的歷史雕琢，一邊記載著劉時棟的生活運作，一邊承載著「物」最初的資訊。那碎片化資訊所構成的終極圖像，從而有了某種集體和個人乃至於環境與生活的詩性紀事特質。

> 劉時棟的作品投影的是那城市裡被棄置的「冗餘」。這些「冗餘」映射出整個當代社會裡那些不曾被劃入結構裡的部分，又或者更為正確地說，這些「冗餘」被劃歸在當代社會結構中那永遠被忽視的角落裡。然而也正是因此，身為一個「切身」的城市探索者，劉時棟才能為我們勾勒、描繪出整體社會結構中那被刻意遺忘與忽視的佈建。更有甚者，這些「冗餘」近一步地構成了當代社會中，那些被「冗餘」的身體與生命的情狀。作品《殘片人》給出了藝術家劉時棟身為臺北這城市的漫遊者，其對臺北都市文化的深刻觀照與省思。

> 如果說劉時棟實踐了一個「藝術實驗」計畫，那麼或許其真正的意涵還在於破壞了「實驗」所追求的「普遍性」理解或事實的本質。亦是在此處，劉時棟的計畫承接了源自於達達以來的反藝術態度從而其創作計畫具備了前衛與當代的意義。（文 | 朱貽安）

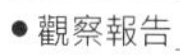

● 觀察報告

朱貽安〈切身而做：劉時棟斷簡、殘片裡的美學探索及省思〉，全文請見
https://mag.clab.org.tw/clabo-article/clab-creators-liu-shih-tung-lands/

表演的政治性—惹內的《陽台》在臺灣
Politics in Performances — Genet's *Le Balcon* in Taiwan

「表演的政治性——惹內的《陽台》在臺灣」將繼續惹內的思維，從《繁花聖母》的性別議題跨到作家後半生關注的政治議題。透過此進駐與創作計畫，野孩子肢體劇場以其擅長的肢體劇場形式，從影像、聲音、裝置等藝術領域，尋找新的跨界合作對象，於進駐期間通過表演者與藝術家的共創，及不同藝術項目與社會民眾互動的創作，共同提問關於表演目前在臺灣文化中所帶來的含義。

This project extends Genet's thought, covering gender issues in *Our Lady of the Flowers* and political issues Genet engaged himself with for the rest of his life. With this project, L'Enfant S. Physical Theatre looks for new partners from other disciplines such as visual art, sound art, and installation art. The Theatre expects to promote discussion on the significance of performing arts in the context of Taiwan's culture by collaborating with performers and artists as well as different forms of works created through the participation of the public.

CREATORS

野孩子肢體劇場
L'Enfant S. Physical Theatre

「野孩子肢體劇場」為肢體默劇表演者姚尚德於 2011 年創立，期望從默劇肢體出發，試圖走出劇場空間，回歸在地文化，以表演尋找與人及社會更深的互動。此外，更期望藉由開辦常態性的肢體工作坊及課程，培養對表演藝術有興趣的弱勢青少年人文修養及表演專長。

L'Enfant S. Physical Theatre was found by YAO Sun-Teck, a corporeal mime artist, in 2011. Starting from corporeal mime, the Theatre attempts to seek in-depth interaction between people and the society by stepping out of theatrical spaces and returning to local cultures. Besides, it aims to cultivate socially vulnerable youth by organizing regular workshops and courses about corporeal mime.

為何而戰
中華民國國家生存

▌野孩子肢體劇場，「表演的政治性—惹內的《陽台》在臺灣」，2018_

L'Enfant S. Physical Theatre, *Politics in Performances- Genet's* Le Balcon *in Taiwan*, 2018_

> 「戲劇如何當代？」是野孩子肢體劇場進駐在 C-LAB 期間一個重要的研究隱題。繼 2018 年 6 月的《繁花聖母》（Notre-Dame des Fleurs）之後，野孩子肢體劇場以《窯．臺》作為對法國劇作家尚．惹內（Jean Genet）的延續性研究，透過對慾望主題的探索進行各種人間視野的探照，無論人們為它蓋上的是善、是惡、是真、是假、是美、是醜、是罪、是罰、是愛、還是恨的遮羞布，透過戲劇將人性最難以直視的部分掀開進行解剖，賭它就是打開潘朵拉盒子的那把鑰匙。在連結批判思考與藝術行動的面向上，野孩子直接與外部導演進行合作，由留法劇場藝術博士王世偉導演親自翻譯惹內的劇作《陽台》（Le Balcon），改編劇本來進行《窯．臺》的劇構，不只讓劇本主題呼應且切合在空總的駐村主題，意圖解構臺灣社會的威權歷史符號，在考察空總的製作條件時，還以特殊場域創作的發展原則來進行場面調度，巧妙地映射空間的場所歷史，向我們展示了一種當代劇場美學在戲劇空間生產上的敏銳與獨特。

> 這次野孩子肢體劇場走進了黑盒子，但卻不同於過去的小型替代空間，也不是一般戲劇演出所使用的純鏡框式舞台，而是空總這座建於 1970 年代的中正堂。更明確地說，在《窯．臺》裡，王世偉導演不僅處理了劇本內部空間的場面調度，也對展演空間選址保持機敏，以提供給觀眾更有現實場域召喚力的戲劇經驗。《窯．臺》改變了傳統戲劇在鏡框式舞台上製造幻象的方式，不被動地接受表演場所狀態，而更積極去尋找與文本內涵具有呼應能量的社會空間形式，以此發展一種能與現實語境對話的情境建構，並替代傳統劇場舞台的戲劇空間系統。

> 《窯．臺》是野孩子的惹內惡之華二部曲，在 C-LAB 的駐村工作團隊與演出的組織模式，保持其一貫昇華嘲弄力量的表演實力，野孩子也透過外部導演的參與來挑戰劇團原本的創作風格，誠如上文的描述分析與詮釋，王世偉導演長期旅法的學術背景確實使《窯．臺》展現一種當代劇場的美學質地。（文｜吳宜樺）

● 觀察報告

吳宜樺〈《窯．臺》：野孩子的惡之華與其超越性別的反父權批判〉，全文請見
https://mag.clab.org.tw/clabo-article/clab-creators-lenfant-s-physical-theatre/

A/V 衝刺班—實驗聲音影像推廣計畫
A/V Intensive Class — Experimental Audiovisual Promoting Project

「A/V 衝刺班 – 實驗聲音影像推廣計畫」結合「噪流」往年的經驗，打造一個推廣交流平台，期進行實驗聲音影像的技術研發、推廣、教學與實踐。除了軟硬體操作的教學之外，也聚焦在創作美學、實驗影音歷史、作品交流、國內外經典藝術作品的賞析與討論，透過建構影音創作的脈絡，使初階創作者有入門學習的機會，同時也使中階創作者重新審視自身脈絡。

Combined with its past experiences, Fluid Noise establishes in this project a platform for the promotion and exchange of research, development, popularization, teaching, and practice in experimental audiovisual art. In addition to the teaching of operating hardware and software, this project also focuses on creative aesthetics, the history of experimental audiovisual art, the exchange of innovative concepts, and the discussion of local and international classic works of art. By establishing the context of audiovisual art, this project allows beginners to learn from the primary level and offers opportunities for intermediate users to re-examine their creation context.

CREATORS

噪流 Fluid Noise

「噪流，匯集每一個吵雜的力量」，匯集每一個河川般的生命，將臺灣的聲音藝術發聲於世界。

「噪流」為失聲祭創辦人姚仲涵於 2011 年所設立的非營利組織，目前單位負責人為影音藝術家葉廷皓。噪流致力於推廣實驗聲音影像藝術，連結國內外創作者與團體，培養創作、評論與執行人才，並策劃相關展演活動，將作品從臺灣推廣至國際。

"Fluid Noise brings together every force of noise." Fluid Noise gathers every stream-like individual and gives voice to Taiwan's sound art to address the world.

Fluid Noise is a non-profit organization founded by YAO Chung-Han, the founder of Lacking Sound Festival. It is currently maintained by audiovisual artist YEH Ting-Hao. Fluid Noise dedicates to promoting experimental audiovisual art, connecting local and international artists and collectives, and cultivating talents in creation, critique, and administration. By organizing relating performances and activities, it aims to promote works of art from Taiwan to the international stage.

▌噪流，「A/V 衝刺班一實驗聲音影像推廣計畫」，2018_

Fluid Noise, *A/V Intensive Class–Experimental Audiovisual Promoting Project*, 2018_

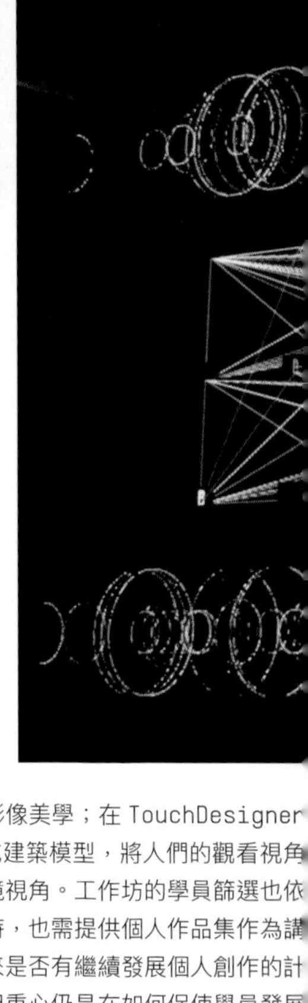

> 「Audiovisual」如其名稱，簡而言之即為影像與聲音結合的藝術，亦稱之為「音像藝術」。但廣義的「音像藝術」亦同時包含了電影、電視、錄像等藝術類型，如查閱國內相關大專院校所開設的「音像藝術研究所」，會發現它們多是針對以電影、電視、錄像等影像為主的創作類型。「噪流」所欲推廣的，則鎖定在以往聲音藝術類型中的「Audiovisual」，且展演形式多為現場演出。

> 「噪流」在進駐計畫的前期，聯合不同的講師群，分別開出 PureData、Processing、MaxMSP、TouchDesigner、3D 模型應用等課程，教授這些軟體在各自影音應用上的強項：在 PureData 的課程上認知聲音的波型，並利用軟體控制波型生成聲音演出中的聲音基底；在 Processing 與 MaxMSP 課程上以圖學影像為基礎，協助已有基礎的學員以演算法發展自身的影像美學；在 TouchDesigner 與 3D 模型應用課程中，則運用當今網路容易取得的現成建築模型，將人們的觀看視角拉入其中，創造彷如可隨意遊走、穿梭建築體的虛擬實境視角。工作坊的學員篩選也依據著「培育創作者」的目標。學員提交工作坊申請的同時，也需提供個人作品集作為講師群的評選依據，講師們則分別針對學員創作特性、未來是否有繼續發展個人創作的計畫等，進行篩選。課程雖貌似以技術作為主要的骨幹，但重心仍是在如何促使學員發展個人的創作，因此也不乏是透過小組甚至一對一教學的方式，統合講師們過往自身遇到的問題，予以學員創作發展上的協助。在半年的進駐計畫中，同時設立了期中小規模演出呈現與最後中型演出交流的目標，一方面讓有潛力的學員在經歷工作坊之後，有將作品完整發表的機會，二來也同時透過演出的舉辦，邀請已有豐富經驗的 Audiovisual 創作者演出，以期達到交流與示範的功能。（文｜馮馨）

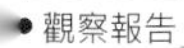

• 觀察報告

馮馨〈從籌辦演出轉往 Audiovisual Art 創作者的培育〉，全文請見
https://mag.clab.org.tw/clabo-article/clab-creators-fluid-noise/

「刷臉」時代的反統治鏈
The Anti-domination in the Era of "Face Swiping"

作為與世界溝通之基本界面「臉」——這個原本人類最特殊的辨識部位，在演算法架構下、被定上座標，提取特徵。成為數據、矩面、群集，甚至是「貨幣」——這個只需遠距離觀測，就能夠獲得的材料進行信息解構，若成為串連所有個人信息的「鑰匙」，會如何影響未來的社會規則？面對不同社會屬性用來解釋科技後行為的詞彙，未來應該會被視為一種文化語言的預備路徑來獨立研究。中國把一切關於人臉的識別技術，用了一個「刷臉」涵蓋——從原本人際文化中對現象反諷（厚顏）的網絡用語，轉換成數據貨幣（支付）。這種對未曾發生的經驗和工具進行詮釋的文化戰爭，表明技術文明的權力架構，已經發生明顯位移。人權價值追求不同的臺灣，又該站在哪一種視域與詮釋技術，來對抗上述這種權力架構？

As the primary interface for communication with the world and the most distinct part of human beings, the "face" is incorporated by a coordinate system for capturing features under the structure of the algorithm. It becomes a data, a moment of area, a cluster, or even a "currency" – materials that can be observed from afar but can provide abundant information once they are analyzed. If the face becomes the "key" to connect all personal information, how would it affect our future social norm? Vocabularies of different social characteristics are utilized to interpret actions behind technology. The "face" will be applied as a preparatory path for cultural language and be studied independently. Chinese government reduces all identification techniques with a brief term "face-swiping," transforming the sarcastic internet slang (of being boldfaced) into digital currency (payment). Such a cultural war which interprets new experiences and tools manifests that the power structure of technological civilization has shifted significantly. What perspectives and interpreting methods should Taiwan's people, who pursue different human values, take to fight against the power structure aforementioned?

CREATORS

施懿珊 SHIH Yi-Shan

空中自體動力宣言（創辦人），主要關注對象：物件化、科技史、材料史、文化批評、文化生態建構、未來人類介面、對話工具開發、（藝術）語言轉譯。自2019年起開始常態參與由虛構研究者——賴火旺發起的「議題串連」。

SHIH Yi-Shan is the founder of Declaration of Airborne Auto-Dynamic Power. She focuses on objectification, history of science and technology, history of materials, cultural criticism, cultural and ecological construction, futural human interface, development of dialogue tools, and (artistic) language translation. In 2019, she started to participate in Issue Connection initiate by a fictional researcher LAI Ho-Wan.

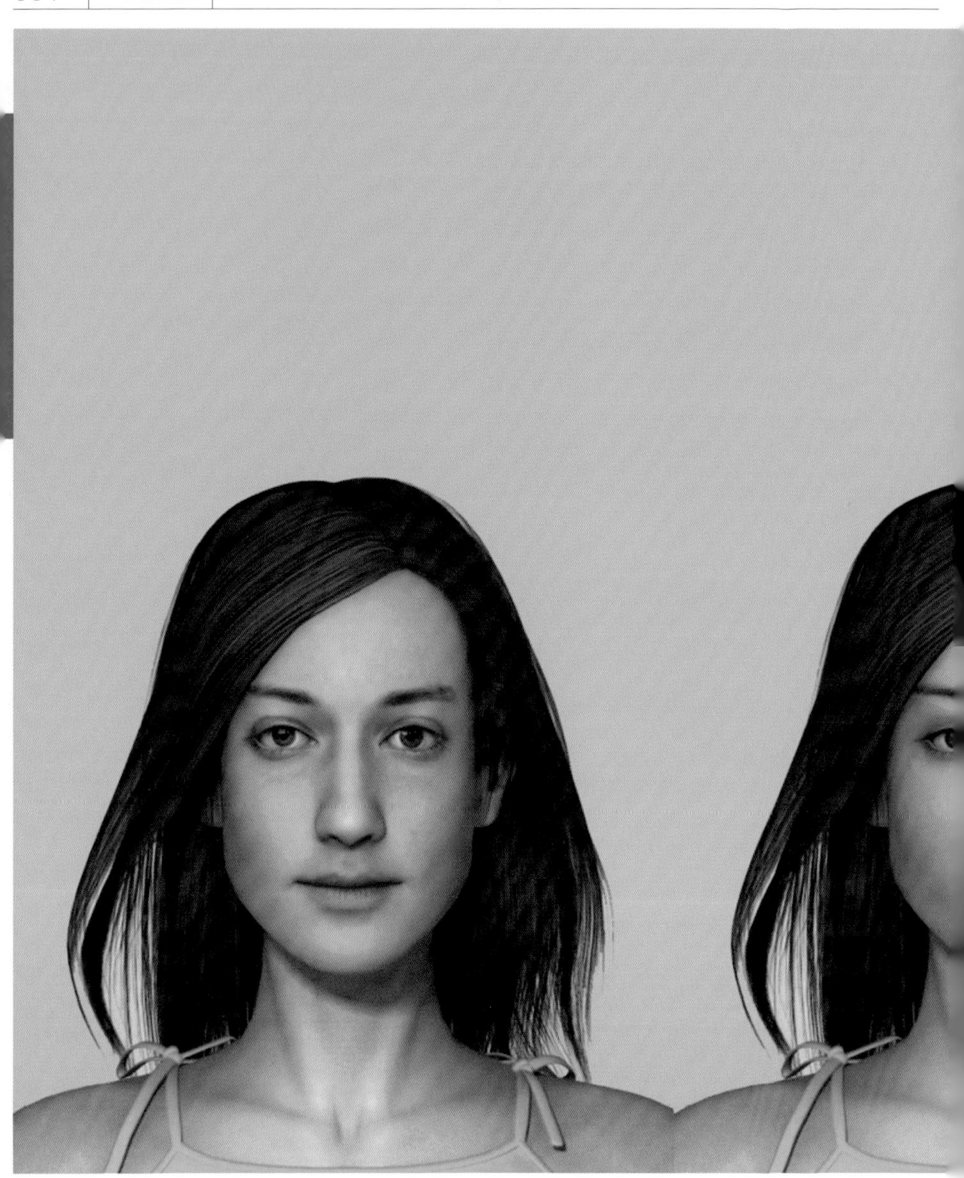

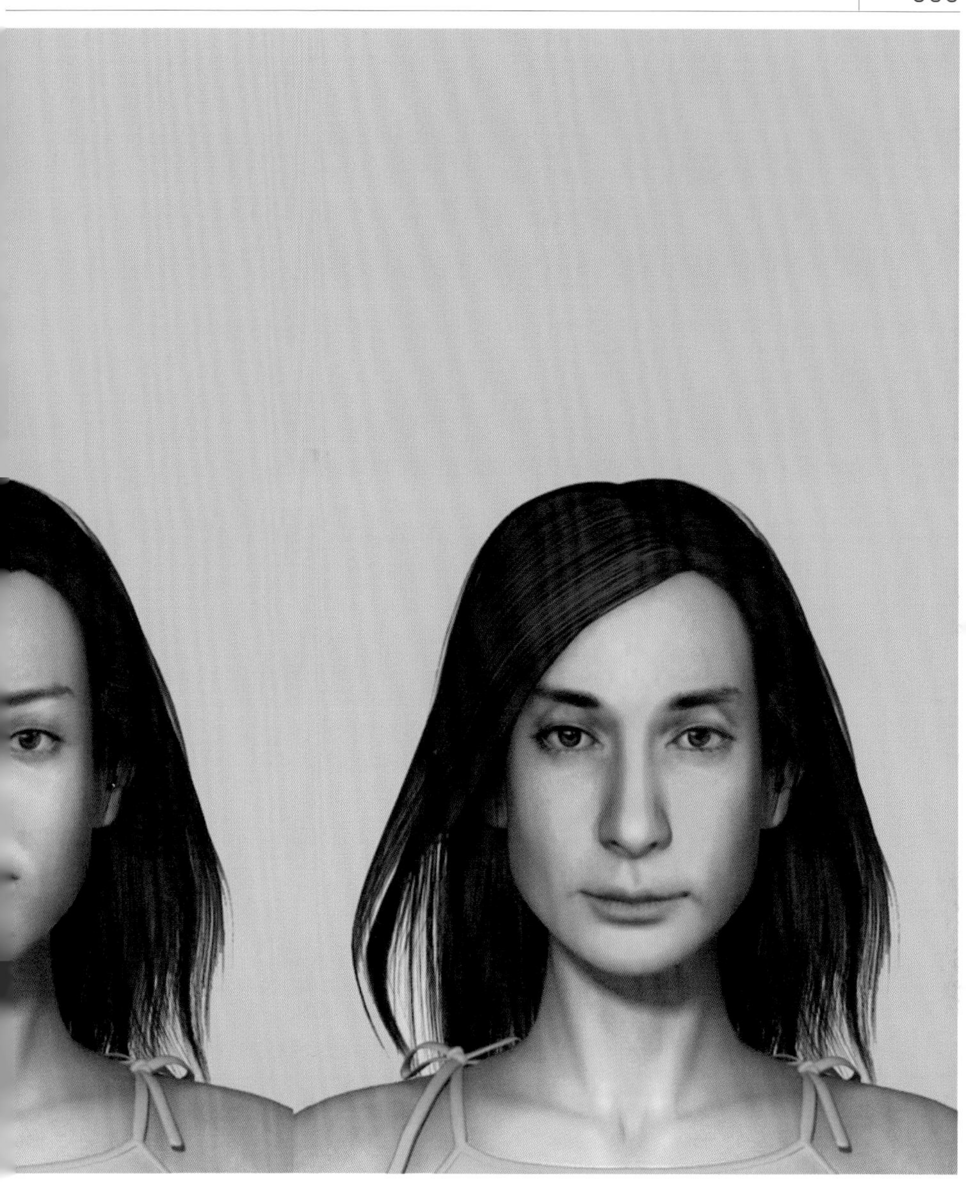

▌施懿珊，「『刷臉』時代的反統治鏈」，2018_

SHIH Yi-Shan, *The Anti-domination in the Era of "Face Swiping"*, 2018_

> 施懿珊於臺灣當代文化實驗場進駐期間，舉辦了多活動。這些活動多為施懿珊邀請與其熟稔，並與其近年進行的藝術和文化實踐相關的學者、專家和創作者進行對談的座談會。由於「『刷臉』時代的反統治鏈」進駐計畫涉及技術門檻極高的網路平台或社群網站的大數據和臉部辨識系統、同樣仰賴大數據、深度學習（deep learning）、人工智慧，或更精確地說，「遞回式神經網絡」（Recurrent Neural Network，RNN）等先端數位科技的無人駕駛自動車，以及該如何從技術的角度，對這些社群網站的臉部辨識系統和無人車進行解構和抵抗的問題，這樣的計畫需要其他跨領域團隊的支援和協助。

> 施懿珊這次的進駐計畫無疑是一次在語義和隱喻層次極具吸引力、在主觀概念層次極具創造力，但在現實的挑戰性和可能性層次卻些許令人失望的嘗試。雖然施懿珊在進駐計畫的執行層面有許多現實上的困難、阻礙和限制，但藝術和文化實驗（踐）畢竟不（僅）是天馬行空的語義和隱喻操作和想像。就對數位（人工智慧）系統進行的扎扎實實地抵抗而言，整個抵抗計畫和策略的研擬或許不能僅停留在「常民文化」、「鄉民文化」或「常民智識」，即「從社會意義的發現、技術邏輯的研擬到技術的破解」的形式性的層次，還必須真真切切地對數位（人工智慧系統）及其獨體、集體和技術邏輯進行更進一步的深入研究，俾以此研擬真正具有效力和創造力的抵抗計畫和策略，進行真正能發揮效能的抵抗性文化實驗（踐）。

> 分層別類，從獨體與獨體在這個數位世界上具有的（階級、階層等）關係、數位服務提供業者對其數位服務或軟體研發方法的研擬、到軟體工程師對軟體進行研發的整個程序，到數位系統中各式各樣的物件，我們看到的是一個得以將現今所有對數位世界的研究架構給涵納進來的數位世界現象學的分析架構。這或許是我們能從施懿珊的 CREATORS 進駐計畫中發現與我們切身相關的最大意義。
（文｜楊成瀚）

● 觀察報告

楊成瀚〈軟體與主權的想像性變遷：論施懿珊 CREATORS 進駐計畫的可能和侷限〉，全文請見 https://mag.clab.org.tw/clabo-article/clab-creators-shih-yi-shan-the-anti-domination-in-the-era-of-face-swiping/

身為藝術工作者，我們如何組織自己？
As an Art Worker, How Do We Organize Ourselves?

此計畫探索的主要三個方向：

一、身為藝術工作者，我們如何在當代的藝術世界中運作？我們如何組織自己？

二、藝術家在國家補助、市場機制，與獨立自給自足之間，還存在著甚麼樣的可能？

三、若共享、共創是藝術家所需要的替代方案，則人類歷史上曾經、或正存在的共享型態，能提供我們何種實踐的經驗？以此建立思考「藝術家身份」時的理論脈絡與基礎，再將焦點置於臺灣處境，探問在臺灣的脈絡中，藝術家們如何組織自己？「藝術家身份」如何被建立？其關注面向將以當代舞蹈為初步範疇，再適時拓展到（廣義的）表演藝術與視覺藝術領域。

This project consisted of three main orientations as follow:

1. As an artist, how should we work in contemporary art world? How should we organize ourselves?
2. Aside from public funding, market mechanism, and self-sustainability, what are other possibilities for artists?
3. If sharing and co-working are alternatives for artists, what kinds of practical experience can previous and current modes of sharing can be offered to artists? These questions serve as the theoretical context and basis for discussing “identities of artists.” We should further focus on Taiwan's context and rephrase these questions as: how do artists organize themselves? how are “identities of artists” constructed? This project takes contemporary dance as its primary target, and then extends to (general) fields of performing and visual art.

CREATORS

吳孟軒 WU Meng-Hsuan

國立臺北藝術大學舞蹈研究所博士生、《PAR 表演藝術雜誌》舞蹈類特約作者。曾任Mekong Cultural Hub策展助理、臺北藝術大學妖山混血盃跨域創意實驗室策展人、《Artplus藝術地圖》雜誌表演藝術類編輯、國藝會表演藝術專案評論人。為舞者、創作者，也為評論人、研究者，關注表演與創作。

WU Meng-Hsuan is a PhD student at Graduate School of Dance Studies, Taipei National University of the Arts, and an accredited writer for *Performing Arts Review*. She served as a curatorial assistant of Mekong Cultural Hub, a curator at the Genie Lab at Taipei National University of the Arts, an editor of performing arats of *Artplus*, and a critic of performing art project for National Culture and Arts Foundation. She is a dancer, creator, critic, and a researcher on performing arts and art creation.

▌吳孟軒，「身為藝術工作者，我們如何組織自己？」，2018_
WU Meng-Hsuan, *As an Art Worker, How Do We Organize Ourselves?*, 2018_

> 吳孟軒注意到，今天在臺灣成為一位藝術工作者意味著必須時時與國家補助，以及全球資本流動結構周旋。一方面，他有可能是以個體的形式直接與藝術機制共處，繼而必須面對各種斜槓人生與非典勞動所帶來的碎裂化、原子化處境。另一方面，他也可能加入各種藝術工作者的聚集形式，意即吳孟軒所說的「共的光譜」——舉凡最基本的藝術團體（早年則是協會、畫會）、劇團和舞團，或者更為複雜的合作社，乃至於社會企業的組織型態。然後發現所有聚集形式的存續，處處受到國家藝文補助的治理邏輯所牽引。簡言之，一位年輕藝術工作者如果要在當前的藝文生態中「活得好」，撰寫各式創作、展演、駐村，甚至是空間營運補助的企劃書，幾乎已是無庸置疑的必備技能。因為唯有掌握與機制之間的這層交換網絡關係，他才有可能在文化資源的羽翼下順利成長。但問題是，我們其實很少認真思考，這裡所謂的「活得好」究竟指的是什麼？

> 如果將 1990 年代解嚴後的藝文環境，與當前這種無論視覺還是表演藝術，都深深嵌入國家補助科層結構的趨勢相互對照，會相當有意思。1990 年代中期，我們尚能看到如後工業藝術祭（破爛生活節）這種藝術家彼此之間毫無目的性的聚集形式，以及在活動過程中，各種機遇性、突發性、純粹耗費式的自主展演。當時這些或靜態或動態的創作生產活動，自然是沒有補助支援的。但也正因如此，它們完全屬於法國哲學家喬治・巴塔耶（Georges Bataille）所說的「耗費性經濟」，而與今天這種任何活動都必須置於經費核銷思維之下的「計畫性經濟」的大環境氛圍大相逕庭。

> 唯有緊扣著「藝術工作者的聚集形式」的課題展開一方面是橫向的案例分析，以及另一方面屬於縱向的系譜學考察，我們或許還有機會在應付機制性疲乏（institutional fatigue）之餘，不至於被其所拖垮，並且在重新想像「活得好究竟意味著什麼？」這件事情上，不至於虛耗過多不必要的力氣。這點，是關於「共」的機制思考之所以非常迫切的根本原因。（文｜王聖閎）

● 觀察報告

王聖閎〈藝術家的聚集，所為何事？側記吳孟軒與其關於「共」的機制思考〉，全文請見
https://mag.clab.org.tw/clabo-article/clab-creators-wu-meng-hsuan-why-artists-gather/

原住民族文化跨藝平台—藝術誌研發計畫
Taiwan Indigenous Interdisciplinary Arts and Culture Connection — Art Magazine Research and Development Project

本計畫集結五位深耕原住民藝術領域之研究、觀察和執行者，組成計畫團隊，重新評估、分析、建議「原住民表演藝術推廣平台」的定位及功能，期望能作為未來表演／視覺藝術平台推廣原住民藝術的諮詢及合作之參據。團隊進駐後，持續透過各地文化田野、深度訪談。進駐期間包括團隊進度會議、活動及展示行動，則開放原住民族的創作者、策展人與民眾進入現場。

This project brings together five members who have been long devoted to research, observation, and practice of indigenous arts; it aims to re-evaluate the position and function of Taiwan Indigenous Performing Arts Connection and gives advice for the platform with close analysis. The result of this project can become the reference for performing/visual art platform which promotes indigenous art in the future. Our team will keep working on fieldworks and in-depth interview during our residence in C-LAB. We welcome public visit by opening meeting, events, actions for public participation, wishing to attract indigenous artists, curators, and citizens to visit our space.

CREATORS

鄒欣寧 TSOU Shin-Ning

國立臺北藝術大學戲劇研究所畢業，主修劇本創作。曾任「莎士比亞的妹妹們的劇團」行政經理、《誠品好讀》、《PAR表演藝術》雜誌編輯，於《誠品好讀》任內與編輯團隊同獲金鼎獎最佳專題製作。現為自由文字工作者，寫作領域含括採訪報導、劇本創作、表演藝術評述等。

TSOU Shin-Ning graduated from the Graduate School of Theater Arts at Taipei National University of Fine Arts and specialized in playwright. She has been working as administration manager of Shakespeare's Wild Sisters Group, the editor of *Eslite Reader* and *Performing Arts Review* and was awarded the best special project prize of the Golden Tripod Awards when she worked for *Eslite Reader*. She is now a freelancer in writing. Her fields of writing include journalism, playwright, and performing art review.

鄒欣寧，「原住民族文化跨藝平台—藝術誌研發計畫」，2018_

TSOU Shin-Ning, *Taiwan Indigenous Interdisciplinary Arts and Culture Connection – Art Magazine Research and Development Project*, 2018_

› 原住民歷史的複雜面向，牽涉到幾度不同政權移轉所造成的遷徙和剝奪，因而「被迫在自己的土地上流浪」，幾乎成為全球各地原住民族的基本處境。1980 年代臺灣原住民運動興起，重新凝望和拾回祖先的生活，各種組織方法開始實驗、逐漸成形，試圖以藝術的能動性見證與處理歷史的演變，延續思考和行動。在此歷史背景下，面對全球化與數位化的浪潮，原民藝術的推廣，首先要面對的任務，反而是在失落的部落裡，重新思考，今日原住民族的界定，該採取何種策略，建立何種網絡平台、何以詮釋。其中自然更觸及到，不同於都市本身的階級結構，而是提供群落的組織和行動方法，以另類方案，擴延原民藝術的深層影響。

› 由於各地方原住民的歷史情境不一，遭受同化的過程和手段不同，在藝術創作中反應的問題和方式也會展現極大差異的力量，以及省思的角度。在這次的研發計畫中，計畫成員以收集故事作為出發，試圖匯聚各方觀點，從喧嘩眾聲裡整理出經個不同的脈絡和參照。其中，藝術誌首先採取了類似派對、集會的形式，將故事彙整我們可見臺灣幾組不同生存方式的藝術家和藝術行政，透過何種角度理解與實驗生活中的藝術實踐。

› 在充滿失憶症的土地上，故事的採集、流傳與變形，進入創作的思考，創造多變的樣貌和意涵，重點不在於民族樣式的確認，而是一種治癒行為，如同藥物的故事是關於如何治癒、如何保持文化命脈和語言的活力。就如同切羅基民族的表演工作者DeLanna Studi所說：「故事讓我們擺脫深淵並且不孤單。」（文｜周伶芝）

• 觀察報告

周伶芝〈我們要說什麼樣的故事？〉，全文請見
https://mag.clab.org.tw/clabo-article/clab-creators-what-kind-of-story-are-we-going-to-tell/

搗亂了所有的創造者

MASHUP all the CREATORS

展期 Date ▶

2019/6/14

|

2019/7/14

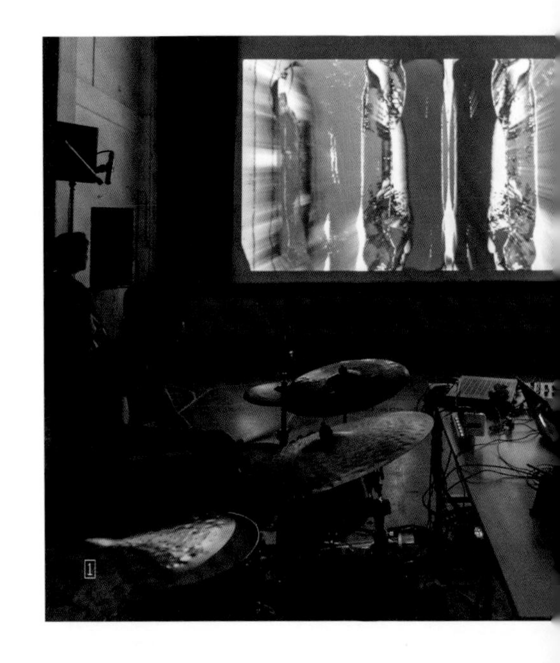

1

2

1 The Art Shelter and Cinema (ASC)、化渡 VedicFormula運算，2018

《Hz》，音像表演
Hz, audiovisual performance

2 吳孟軒，WU Meng-Hsuan，2018_

《身為藝術工作者，我們如何組織自己？》
As an Art Worker, How Do We Organize Ourselves?

3 野孩子肢體劇場　L’Enfant S. Physical Theatre，2018_

《表演的政治性—惹內的《陽台》在臺灣》
Politics in Performances— Genet’s Le Balcon in Taiwan

4「搗亂了所有的創造者」現場　MASHUP all the CREATORS

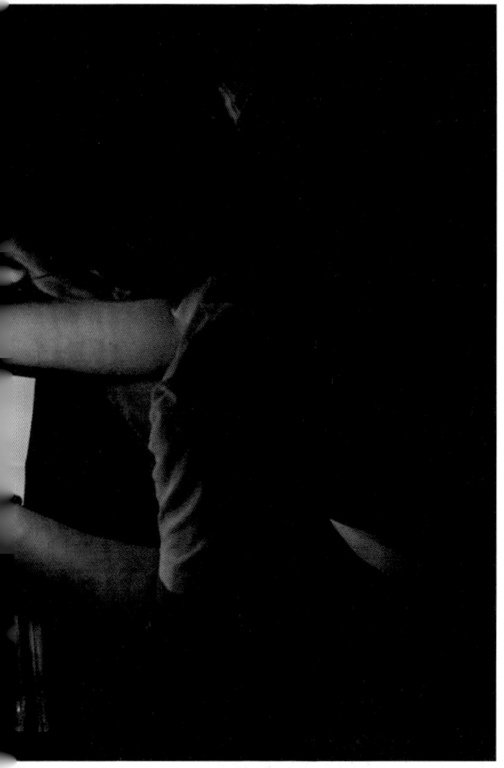

3

4

1 謝杰廷、鍾玉鳳、黃思農、鄭傑 HSIEH Chieh-Ting, CHUNG Yu-feng, Snow HUANG, CHENG, Chieh-Wen，2018_
《譯譜者》，裝置、表演
Transnotators, installation, performance

2 燧人氏 Zuirens，2018_
《Audiovisualizer in Da House.》
Audiovisualizer in da house.

藝術家
ARTISTS

ASC The Art Shelter and Cinema、劉時棟、空氣結構、黃偉軒、謝杰廷、在地實驗、燧人氏、野孩子肢體劇場、吳孟軒、圓劇團

ART SHELTER AND CINEMA, LIU Shih-Tung, Air Structure lab, HUANG Wei-Hsuan, HSIEH Chieh-Ting, ET@T, Zuirens, L'Enfant S. Physical Theatre, WU Meng-Hsuan, Thunar Circus

Hidden Layer of City [臺北版]
Hidden Layer of City [Taipei ver.]

在世界快速數位化與智慧化的發展之下，城市的面貌也由可見的技術景觀轉換進入不可見的監控與運算世界，過去依賴身體感官（如聽覺、視覺等）所認知的現象與建立的連結，已無法應對。而透過科技與微感測裝置的輔助，似乎讓我們取得關鍵的機會，進入由巨量微感測運算與數據調控所形成的數位世界 —— 一個身體感官所無法觸及（無所感）的世界。

數位化城市的運作，大量依賴於各項數據監測與運算調控，這些數據資料背後所代表的現象，如層層隱藏不可見的 layers，交疊在城市之中，形成我們周圍的樣貌和狀態。本計畫使用多項微感測器，例如氣體（CH4, CO, CH-CH3, H2, CH3-CH2-OH, CO2, CO）偵測感測器、土壤物質感測器、植物生態感測器進行實驗，將紀錄的數據 （包含數值、位置、時間等），透過演算法、處理，以轉換成視覺、聲音裝置的方式，進行對於城市中不可見 layers 再現的探測、實驗。

With the rapid development of digitalization and implementation of artificial intelligence, the visible, technical sceneries of urban landscapes are transformed into an invisible world of surveillance and algorithm. Our previous cognition and connection with the outer world which relied on bodily sensations (such as hearing and vision) are no longer sufficient. Technology and microsensors seem to offer key opportunities for us to enter the digital world of big data microsensors and data modulation, a world which is unreachable (un-sensible) to physical sensations.

The operation of a digital city relies heavily on various data monitoring and mathematical modulation. The phenomena behind these data are like invisible layers, overlapping in the city and forming our surroundings. In Hidden layer of City [Taipei ver.], the artist conducts experiments with several microsensors such as gas (CH4, CO, CHCH3, H2, CH3-CH2-OH, CO2, CO) sensors, soil material sensors, and plant ecology sensors. The documented data (such as numerical value, location, and time) are processed through algorithms, surveying and experimenting the invisible layers in the city through visual installations and sound installations.

CREATORS

張永達 CHANG Yung-Ta

1981 年生於臺灣，目前工作與居住於臺北。作品包含多種形式，如聲音—影像、實驗聲響、聲音裝置和現場演出。關注與善於觀察日常生活中的細微變化、易於忽略的物理現象和聲響，以空間裝置呈現，意圖在重度視覺導向的現今環境之下，喚起用身體其他感官去觸診與開發這個世界。近年與舞蹈、音樂其他領域的藝術家合作，並於亞洲、歐洲、北美和南美洲多國聯展和藝術節展演。

Born in Taiwan in 1981, CHANG Yung-Ta is autodidact and residing in Taipei currently. His works take on a variety of forms such as audiovisual, experimental sound, installations and live performances. He focuses on and is keen in observing tiny changes in daily life and physical phenomenon and sound that are easily ignored. He then presented them as spacial installations. Intending to arouse other body senses to palpate and exploit the world in a such highly visual-oriented environment. He also collaborated with choreographers, musicians, and artists in different fields. His works have been featured in numerous collaborated shows and festivals in Asia, Europe, Australia, North America, and South America.

Max Console

Filter

SCD30 • SCD30 co2(ppm): 610 temp(C): 28.599602 humidity(%): 39.334106
SCD30 • SCD30 co2(ppm): 602 temp(C): 28.58625 humidity(%): 39.382935
SCD30 • SCD30 co2(ppm): 588 temp(C): 28.612953 humidity(%): 39.311218
SCD30 • SCD30 co2(ppm): 583 temp(C): 28.642326 humidity(%): 39.277649
SCD30 • SCD30 co2(ppm): 585 temp(C): 28.655678 humidity(%): 39.187622
SCD30 • SCD30 co2(ppm): 584 temp(C): 28.669029 humidity(%): 39.201355
SCD30 • SCD30 co2(ppm): 589 temp(C): 28.655678 humidity(%): 39.122009
SCD30 • SCD30 co2(ppm): 592 temp(C): 28.655678 humidity(%): 39.109802
SCD30 • SCD30 co2(ppm): 591 temp(C): 28.669029 humidity(%): 39.137268
SCD30 • SCD30 co2(ppm): 594 temp(C): 28.685051 humidity(%): 39.112854
SCD30 • SCD30 co2(ppm): 591 temp(C): 28.685051 humidity(%): 39.086914
SCD30 • SCD30 co2(ppm): 602 temp(C): 28.669029 humidity(%): 39.123535
SCD30 • SCD30 co2(ppm): 611 temp(C): 28.628975 humidity(%): 39.068604
SCD30 • SCD30 co2(ppm): 612 temp(C): 28.540855 humidity(%): 39.163208
SCD30 • SCD30 co2(ppm): 606 temp(C): 28.527504 humidity(%): 39.250183
SCD30 • SCD30 co2(ppm): 598 temp(C): 28.58625 humidity(%): 39.292908
SCD30 • SCD30 co2(ppm): 590 temp(C): 28.58625 humidity(%): 39.306641
SCD30 • SCD30 co2(ppm): 583 temp(C): 28.628975 humidity(%): 39.311218
SCD30 • SCD30 co2(ppm): 590 temp(C): 28.628975 humidity(%): 39.402771
SCD30 • SCD30 co2(ppm): 589 temp(C): 28.655678 humidity(%): 39.379883
SCD30 • SCD30 co2(ppm): 580 temp(C): 28.711754 humidity(%): 39.360046
SCD30 • SCD30 co2(ppm): 585 temp(C): 28.696402 humidity(%): 39.204407
SCD30 • SCD30 co2(ppm): 592 temp(C): 28.711754 humidity(%): 39.128113
SCD30 • SCD30 co2(ppm): 604 temp(C): 28.696402 humidity(%): 39.0625
SCD30 • SCD30 co2(ppm): 599 temp(C): 28.711754 humidity(%): 39.053345
SCD30 • SCD30 co2(ppm): 592 temp(C): 28.711754 humidity(%): 39.027405
SCD30 • SCD30 co2(ppm): 589 temp(C): 28.783859 humidity(%): 39.045715
SCD30 • SCD30 co2(ppm): 590 temp(C): 28.767838 humidity(%): 39.007568
SCD30 • SCD30 co2(ppm): 590 temp(C): 28.797211 humidity(%): 38.972473
SCD30 • SCD30 co2(ppm): 594 temp(C): 28.797211 humidity(%): 38.95874
SCD30 • SCD30 co2(ppm): 606 temp(C): 28.767838 humidity(%): 38.967896
SCD30 • SCD30 co2(ppm): 613 temp(C): 28.783859 humidity(%): 38.90686
SCD30 • SCD30 co2(ppm): 630 temp(C): 28.696402 humidity(%): 38.908386
SCD30 • SCD30 co2(ppm): 625 temp(C): 28.655678 humidity(%): 39.009094

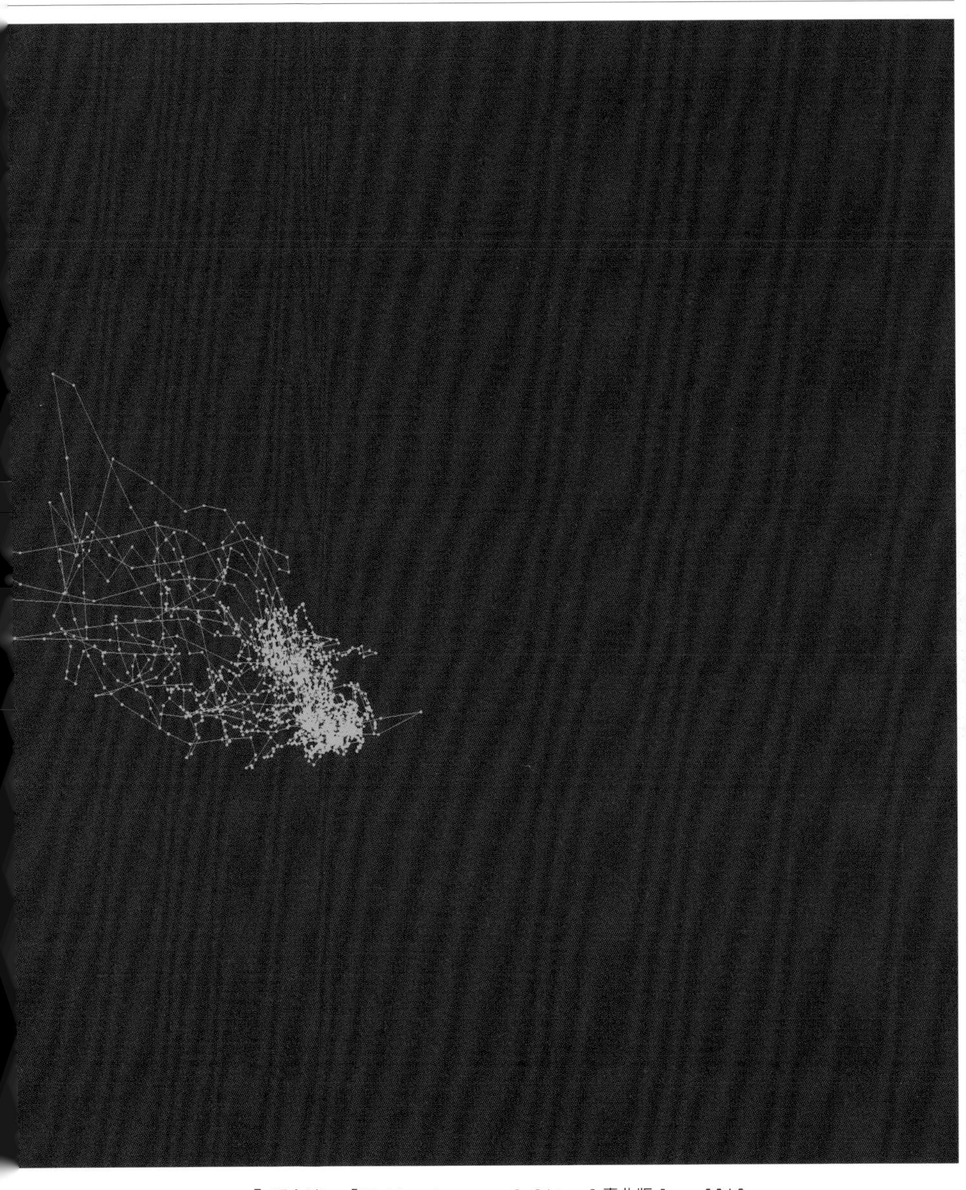

▌張永達，「Hidden Layer of City [臺北版]」，2019_
CHANG Yung-Ta, *Hidden Layer of City [Taipei ver.]*, 2019_

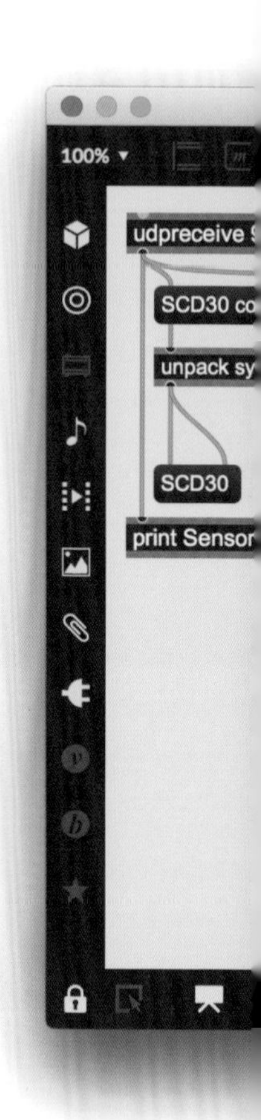

> 將數據透過藝術轉化、轉譯而使其「現形」、「現聲」向來是張永達的創作語彙，在「Hidden Layer of City [臺北版] 」計畫中，更立基於此基礎上並帶入與環境有關的思考。靈感來源除了近年來的排碳量議題，也來自於電影《銀翼殺手 2049》（Blade Runner 2049）或影集《核爆家園》（Chernobyl）中的背景設定，進而選擇二氧化碳為主要關注標的，並發展出相關的創作實踐。張永達從思考都市景觀和二氧化碳之間的關係出發，把具自動濕濕度校正功能的二氧化碳偵測模組結合 GPS 定位裝置，再將所測得的數據，轉化為標記在地圖上實際位置的視覺圖像。

> 張永達試圖藉計畫提問「人類如何能夠去感知到數據的龐雜，感覺到其無限性是超越我們的想像？」就二氧化碳的偵測數據來說，螢幕上所看到從 600ppm 到 1,000ppm 的數值變化是理所當然的呈現形式，他提到：「但若思考介面所呈顯的數據之間到底存在多大的距離，我們對此的認知則是很
糊的，也難以捕捉或感知這在環境裡面改變的實際狀態。」對於人類來說，數據相
於光或者聲音是較不會有直接地感官衝擊，這不僅指向藝術創作的範疇，張永達認
人們在生活中也處於這樣的制約狀態，也就是說，數據早已透過人類習慣的方式在
常裡達成了它的隱匿。

> 藝術創作所追求的貼近事實狀態或真相，廣義而言，與科學驗證的最終目標有相符之處，不過兩
的探究路徑相當不同。「科學家是透過直接的數據赤裸裸地去驗證並得出結果；藝術家則是在人
與感性層面進行轉化，其追求的真理也不指向明確的是非對錯。」對張永達來說，此次進駐過程
未來創作計畫的前置研究跟實驗階段——要如何跳脫以人為中心的思考？如何突破數據轉譯的
段，去想像其他創作上的可能，並藉以關注數據背後那不可掌握性、或者「終究需要回歸到某種
質」，皆是他給予自己的下一道課題。（文｜黃鈴珺）

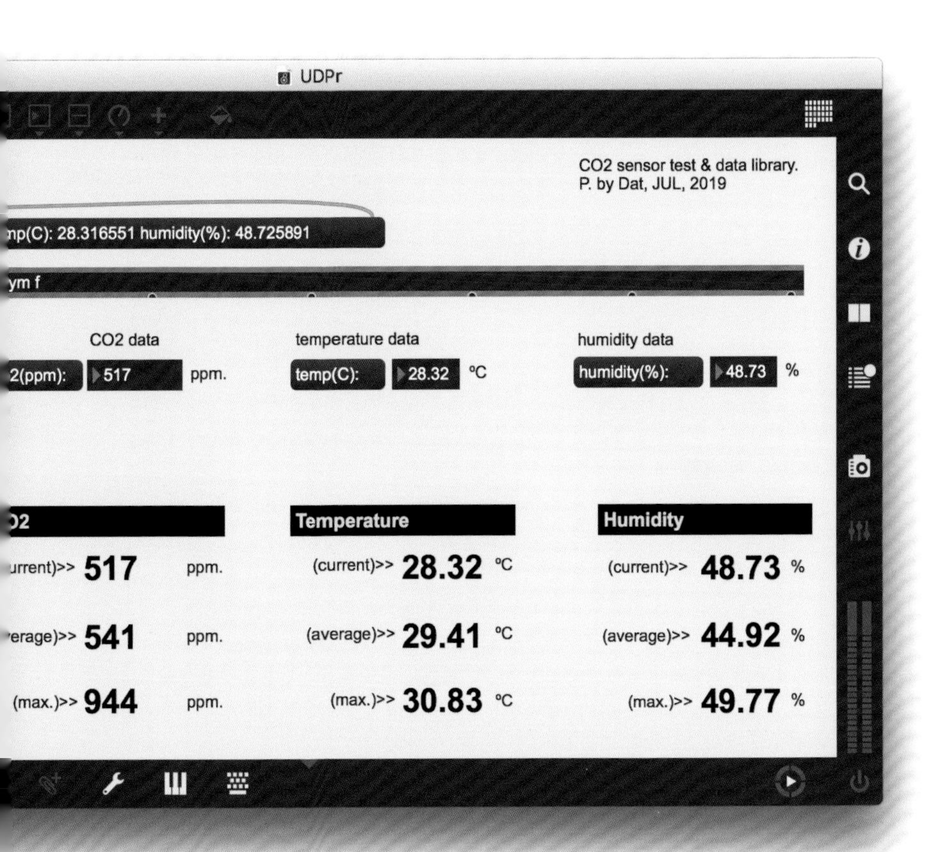

● 觀察報告

黃鈴珺〈數據生態中自然法則的無法則，談張永達的「Hidden Layer of City [臺北版]〉，全文請見 https://mag.clab.org.tw/clabo-article/hidden-layer-of-city-taipei-ver/

Relight
——消逝與正消逝的臺北聚落光實驗計畫
Relight — The Vanished and Vanishing Settlements in Taipei

平日隱身於地景之下的「臺北地下道」，曾經肩負城市人們穿梭移動的重責。然而歷經交通與空間的變革，近年來多數地下道不僅使用率低落，老舊的建築亦常引發安全爭議。自 2016 年起，臺北市地下道已大幅度拆除，迄今僅剩下 37 座，同時逐年持續拆除。經創作者親身走訪與調查發現，看似僅供「通行」用處的地下道，其實與周邊城市空間的發展有重要的共生關聯。而創作團隊又會如何運用「光」，與這個即將消失的場域產生的對話與關係呢？

The underpasses of Taipei once carried the weighty responsibility of seeing to people passing through the city. However, with changes in transportation and spatial arrangement and fewer people using the underpasses, old structures became a safety hazard. Since 2016, large percentages of underpasses in Taipei City have been dismantled. Only 37 underpasses remain now and the number continues to decrease as years go by. By investigating and visiting these sites in person, the artists realize that these underground "passages" are intimately connected with urban developments of nearby areas. Treating light as their medium, how would the team bring about conversation and interaction with these disappearing spaces?

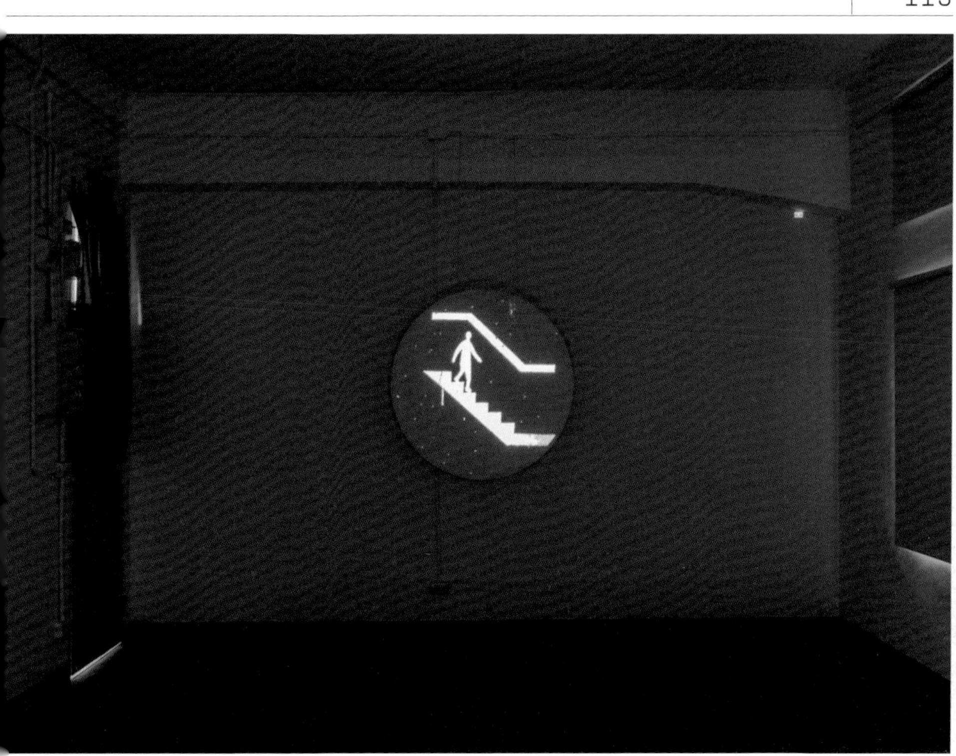

CREATORS

莊知恆 CHUANG Chih-Heng

981 年生於臺北，劇場燈光設計師，畢業於國立臺北藝術大學劇場設計學系碩士班，求學期間致力於鑽研燈光設計領域，擅長以不同觀點切入各式作品，創造出個人獨特風格的光影語彙，近年來經常與臺灣新生代創作者跨界創作；2015 年以燈光設計作品《看得見的城市，看不見的人》代表臺灣赴捷克共和國「布拉格劇場設計四年展」展出，2017 年則以此作品獲得「2017 世界劇場設計展專業燈光設計組」銅獎；2018 年與知名音樂人林強共同創作聲光體驗作品《神木約定》，於 018 年嘉義臺灣燈會展出。

Born in 1989, Taipei, CHUANG works as a theatrical lighting designer. Graduated from Department of Theatrical Design and Technology, Taipei National University of the Arts, CHUANG conducted researches on lighting design during his years at school. He employed different perspectives while dealing with various art works, thus creating a unique fashion of manipulating lights and shadows. Collaborating with Taiwanese artists of young generation in recent years, CHUANG represented Taiwan in attending The Prague Quadrennial of Performance Design and Space with his lighting design *Visible & Invisible* in 2015. In 2017, participating the World Stage Design 2017 Exhibition, he won the Bronze Award at the Professional Lighting Designer with the same piece. In 2018, he co-created the immersive light installation *Tree of Light/Tree of Life* with Taiwanese musician, Lim Giong, and the piece was exhibited on the 2018 Taiwan Lantern Festival in Chiayi.

附近地標｜好市多北投店、關渡捷運站
附近地標｜和信醫院
大同區-重慶民權地下道
大同區
大同區-啟聰地下道
連接路線｜重慶北路三段
同路線是否有斑馬線｜有（相對方便、有行人號誌燈）
附近地標｜啟聰學校
大同區-重慶民權地下道
萬華區-萬大富民地下道
萬華區
萬華區-萬大富民地下道

▌莊知恆，「Relight — 消逝與正消逝的臺北聚落光實驗計畫」，2019_
CHUANG Chih-Heng, *Relight — The Vanished and Vanishing Settlements in Taipei*, 2019_

> 原為劇場燈光設計師的莊知恆，近年開始從黑盒子中出走，涉足視覺與裝置藝術領域。從「光」這一並無實體的物理媒介出發，關注都市更新議題中，那些即將消逝的都市空間及其記憶是否能有再次活絡的可能。「Relight—消逝與正消逝的臺北聚落光實驗計畫」（簡稱「Relight」）著眼於臺北市內逐漸被行人遺忘的地下道。在藝術家親自踏查後，試圖以其最擅長的媒材——光，作為都市中正消逝空間重新得到關注的契機。

> 「Relight」中將照亮對象指向了已被遺忘的地下道，顯然是採取了全然相反的做法。莊知恆借用了美國建築師馬修．佛德烈克（Mattew Frederick）在《建築的法則》中對於建築空間形式中的「正空間」（positive space）與「負空間」（negative space）說法來指認地下道的懸置狀態。在原書中，形容繪製建築圖時，得以圍塑出形狀的空間稱之為「正空間」，而無法形塑的稱之為「負空間」。一般來說，人們穿越「負空間」，在「正空間」停駐。亦即，「正空間」是目標空間，人們於此流連社交與活動，而「負空間」則是讓人移動的非停駐之過渡性空間。然而，對許多特定群體的人來說，我們認知的負空間，卻有可能成為他者的正空間。（文｜謝鎮逸）

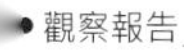
觀察報告

謝鎮逸〈光照亮了誰？莊知恆的光實驗計畫與未知的暗〉，全文請見
https://mag.clab.org.tw/clabo-article/chuang-chih-heng-relight-project/

熒惑蟲計畫
Mars BUG

「察剛氣以處熒惑。熒惑為勃亂，殘賊、疾、喪、饑、兵。」
——《史記．天官書》

火星，是人類近年來發展星際移民很常討論到的星球，同時在古代神話與占星學中都代表了毀滅的能量，在這次的計畫中，火星象徵著城市文明全面且具侵略性的發展。

BUG — Build（創建）— Uprise（高峰）— Garbage（廢棄物）。

本計畫以 B、U、G 三個階段的公開展演，構建近未來逐漸崩毀的城市。以臺灣青年移居海外工作為主題進行資料搜集和文本創作，其中以勞動和遷移的角度，探討人類用勞動建構城市，城市引發遷移，而勞動者在遷移的循環中，身體與思想全然地失去了方向。

"The planet Mars has masculine quality. Mars brings upheavals and turmoil, cruelty and thieving, sickness, death, hunger, and armies."
– *Shiji* [*Records of the Grand Historian*]

The planet Mars has appeared frequently in recent discussions on interstellar immigration. It is also a planet known for its destructive forces in ancient mythology and astrology. This project symbols the destructive, aggressive development of urban civilization.

BUG – Build – Uprise – Garbage.

This project is presented with three stages, including B, U, and G, constructing a future city that is gradually collapsing. Focusing on young Taiwanese individuals living abroad, this project conducts data collection and literature creativity, exploring how human being construct cities through labor and migration. While cities bring forth migration, it also creates a disorientating effect on the bodies and minds of laborers in the process of relocation.

CREATORS

僻室 House Peace

創立於 2018 年，由一群擁有劇場各領域專業的成員組成，範疇包含導演、表演、舞台、燈光、服裝等劇場展演及各類視覺設計。僻室以多元創意為主導，從劇場出發，嘗試與各式創作媒材結合，目標成為共融劇場表演、空間景觀及原創文本的計畫實驗室。主要成員有吳子敬、吳璟賢、吳峽寧、吳靜依、蔡傳仁、陳則妤、劉農、羅宥倫、魏聆琄。

Established in 2018, House Peace was founded by a group of professional theatre workers with specialties in directing, acting, set design, lighting design, costume design, and different kinds of visual art. We are an art laboratory that combines theatre and various elements for exhibitions and productions. Focusing on creativity, House Peace intends to combine theater with different creative media in order to become a lab for inclusive theater performance, experimental spaces, and original texts. Active members of House Peace include WU Zi-Jing, WU Jing-Xian, WU Xia-Ning, WU Jing-Yi, CAI Chuan-Ren, CHEN Ze-Yu, LIU Nong, LUO You-Lun, and WEI Ling-Juan.

▌僻室，「熒惑蟲計畫」，2019_

House Peace, *Mars BUG*, 2019_

- 「熒惑蟲計畫」的啟動源自於劇作家陳弘洋在澳洲打工度假的移動經驗。以火星作為一個「這麼近那麼遠」的地理指涉：這麼近，是火星有水源、有生命跡象、有移居可能；離地球最近，彷彿只要再努力一點就可以到得了的地方。那麼遠，是縱使一切看起來都是如此有可能，卻遲遲都無法落實所有對火星的慾望與想像。

- 全計畫共分為三階段：第一部「漫遊者」為靜態展覽，假設了人類殖民並移居火星以後卻又開始遺棄，僅剩廢棄垃圾囤積的廢墟化場景。第二部《不在這裡》由 2018 年臺北藝穗節首獎作品《我好揪節——你想要的都不在這裡》作為延伸，以劇作家陳弘洋到澳洲打工的心路歷程，探索移居和遷徙如何影響自身的生命情境。現場除了一群表演者扮演的機器人，各自述說火星移民的種種兼具報導又紀錄式的故事，劇作家也親自出演，大段獨白如同一篇感性且飽含世代反思的個人宣言。演出最後以對舞台陳設與場景的拆解與肆意破壞作結，並預告了最終部——一部完整劇場演出《火星》的降臨。劇中透過多個角色人物來說一個有關離家、移民的故事，如何在多種意外與艱難中，親手斷送烏托邦的理想情境。

- 《不在這裡》與《火星》雖然從腳本中牽出了跨國移動與勞動命題，但乍看之下，感性主體並非大於社會性，而直接是一種「感性主體作為社會性」。意即，我們從劇中各個人物身上看到他們都是天底下同時代、同氛圍、同命運的受害者。那些繁複的對話、冗長獨白、青澀歌聲還有一旦哀愁就有眼淚的敢愛敢恨時代，都直接代理了深刻與哲思、歷史與理論。「# 感傷」，成了這趟感性之旅毫無意外的必然產物。《不在這裡》中的毀敗、《火星》的悲劇陰影，以及兩者同樣提及理想與憧憬的泯滅，都不約而同把觀眾導入到集體傷感氛圍的浸潤之中。（文｜謝鎮逸）

● 觀察報告

謝鎮逸〈# 地球很危險：「僻室」登陸火星的廢物、拔克與災難〉，全文請見
https://mag.clab.org.tw/clabo-article/house-peace-mars-bug/

城市塵埃——實踐街 1 號登月計畫
Shijian st. 1 Landing Program

「城市塵埃——實踐街 1 號登月計畫」以位在臺北市北投區實踐街 1 號的中央通訊社員工宿舍荒廢閒置空間為計畫創作與研究方向。「中央社宿舍」是國民政府為在 1949 年隨中華民國政府來臺灣的中央社員工眷屬所建設，許多中央社宿舍位在臺北精華地段，如實踐街、承德路，其房舍現已破舊不堪、閒置超過 19 年。目前中央社宿舍已收回國有財產署所有，北投區的地方文化人士與立委都曾關切中央社宿舍議題，並期盼文化部能活化再造，成為文化空間或藝術聚落。此計畫期望能在官方資源尚未能即時投入前，先行透過藝術創作、展演、研究與進駐等開放方式，帶動官方單位與地方文化組織未來在資源整合、創造閒置空間活化與地方再生契機的多重想像，另一方面也嘗試將相關討論帶入 C-LAB，藉此思考在 1990 年代以後，創作者自主進駐城市角落空間的另一種可能性。

Shijian st. 1 Landing Program is based on the abandoned staff dormitory of Central News Agency (CNA) located in No. 1, Shijian St., Beitou Dist., Taipei City. The CNA dormitory was an accommodation built for staff members of CNA when the KMT government arrived in Taiwan in 1949. Many CNA dormitories were built in supreme locations across Taipei, such as Shijian Street and Chengde Road. The building is now old and worn out and has been abandoned for more than 19 years. The CNA dormitory is now owned by the National Property Administration and has become a focus for cultural workers and members of the Legislative Yuan in Beitou area, hoping that the Ministry of Culture would reconstruct the dormitory into a cultural space or artist village. Through artistic creations, exhibitions and performances, research, and residencies, this project hopes to encourage government and local cultural organizations to envision ways of integrating resources, invigorating unused spaces, and revitalizing local communities. Also, this project intends to bring the discussions into C-LAB about the possibilities of creative workers to autonomously enter spaces peripheral to city spaces.

CREATORS

郭奕臣 KUO I-Chen

郭奕臣的作品不斷透過不同的媒材型式，創造出一種詩意兼具情境式的獨特語彙，作品的核心關注於環境與內心歸屬感的消逝與飄盪的精神狀態，並透過不存在的狀態去顯現對生命本質的探索，作品《入侵》曾代表臺灣參加 2005 威尼斯雙年展臺灣館，並為歷年來參展最年輕的藝術家。

KUO I-Chen uses different media to create poetic and immersive language in his works. He manages to explore the essence of life by presenting the “absence.” In 2005, Kuo brought his *Invade* to Venice Biennial and became the youngest artist ever to represent Taiwan in the exhibition.

林怡秀 LIN Yi-Hsiu

現為影像與藝術評論、文字創作者，長期擔任當代藝術、影像等評論類雜誌編輯、特約企劃及作者，現為自由評論人、文字工作者。

LIN Yi-Hsiu is a visual image and art critic and writer. She has served as an editor for magazines about contemporary art and visual critic, special project planner, and writer for a long time.

郭奕臣、林怡秀，「城市塵埃——實踐街 1 號登月計畫」，2019_
KUO I-Chen, LIN Yi-Hsiu, *Shijian st. 1 Landing Program*, 2019_

「城市塵埃——實踐街 1 號登月計畫」（簡稱「城市塵埃」）的構思源於郭奕臣和林怡秀對臺灣藝文環境的探究與自身經驗，反映在藝術家創作、空間經營、藝術進駐實踐，以及藝文政策等的發展變遷。藉由兩條路徑鋪展其關注範疇的討論，帶出 1990 年代至今的藝文生態面面觀，亦企圖透過場域實踐串起民間與官方的資源，開啟目前官方治理框架外尚缺的藝術動能：郭奕臣以創作為方法，試圖由展覽打開對話；林怡秀則透過報刊呈現的資料研究與座談舉辦做為回應。

「城市塵埃」選擇已荒廢的中央通訊社宿舍位於臺北市北投區的舊址：實踐街 1 號為起點，嘗試在藝術實踐、官方政策以及土地所有者三種立場之間取得平衡點，透過官方資源的挹注，使空間有機會活化為長期性的藝術家工作室，從而發展成具延續性的藝術聚落。

郭奕臣曾提及計畫的初衷在於「探討一個還沒被探討或者不在官方規則裡面的狀態，並藉此突破既有框架。」弔詭的是，「城市塵埃」雖獲得官方資源補助，但是，當計畫實際碰觸到官方結構的癥結問題後，卻無法有結構性地推進。兩人暫時從不得而入的實體空間轉換到紙上與數位空間，包括《登月者通訊》以及未來其他可能的紀錄或出版形式，延續計畫關注的討論。（文 | 黃鈴珺）

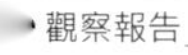

觀察報告

黃鈴珺〈還能如何延續「那個空間」？郭奕臣、林怡秀「實踐街 1 號登月計畫」〉，全文請見 https://mag.clab.org.tw/clabo-article/shijian-st-1-landing-program/

當代城市採集——自己家的雜草茶
Urban Gatherer — Making Your Own Weed Tea

因為喜歡喝青草茶而跟著草藥師上山採草
因為看見了深山裡的除草劑而決定為土地煮一鍋茶
因為雜草們，開始發現人與土地的秘密……

從白皙胖嫩的平凡上班族
到吃草吃土的採集藝術家
療癒自己、療癒人心、療癒地球的真實故事

The love for herbal tea turned into footsteps that followed herbalists into the mountains.
The sight of herbicides in the mountains led to the decision to brew a pot of tea for the land.
While the weeds triggered the mysteries between people and the land...

From a plump, bright-skinned, ordinary office worker.
To a collection artist who meddles with dirt and grass.
This is a true story that heals the self, the mind, and the world.

CREATORS

雜草稍慢/林芝宇
Weed Day/Zo LIN

原從事平面設計，辭職後展開環島旅程並在途中打工換宿，行旅中認識到被拔除的雜草是可治療感冒的藥材「土香」，又遇到教導她採集的青草茶師傅，便開始投入雜草的採集與體驗，並於 2014 年成立「雜草稍慢」進行創作與雜草茶分享。

近年林芝宇也行旅世界各地親試當地植物，更深入認識雜草亦藥亦毒的特性。對她而言，每片土地都有自己的草相，煮出來的茶自然也帶有無法被復刻的獨特風味。從製作到煮茶的各個處理階段，雜草的香氣與觸感會不斷地變化；在由熱變冷的品茶過程裡，味道也一直在流轉——那是一種活的氣味，療癒的五感體驗。

Zo LIN used to be a graphic designer. After her resignation, she travelled around Taiwan and worked in various places in exchange for a night's accommodation. She met an experienced herbalist on the road, who taught how and which plants and weeds to forage. In order to share this acquired knowledge, Lin founded Weed Day in 2014.

In recent years, Lin has begun travelling internationally to taste and test medicinal properties and toxicities of different indigenous plants. She thinks every place has its distinct breed of weeds, which can be is brewed in tea with unique flavor. From picking to preparing the plants, mixing to brewing the tea, the rising aromas and textures are constantly changing—even the process of drinking is one of lively transformation, as each of the five senses is affected differently as the tea moves from hot to cold.

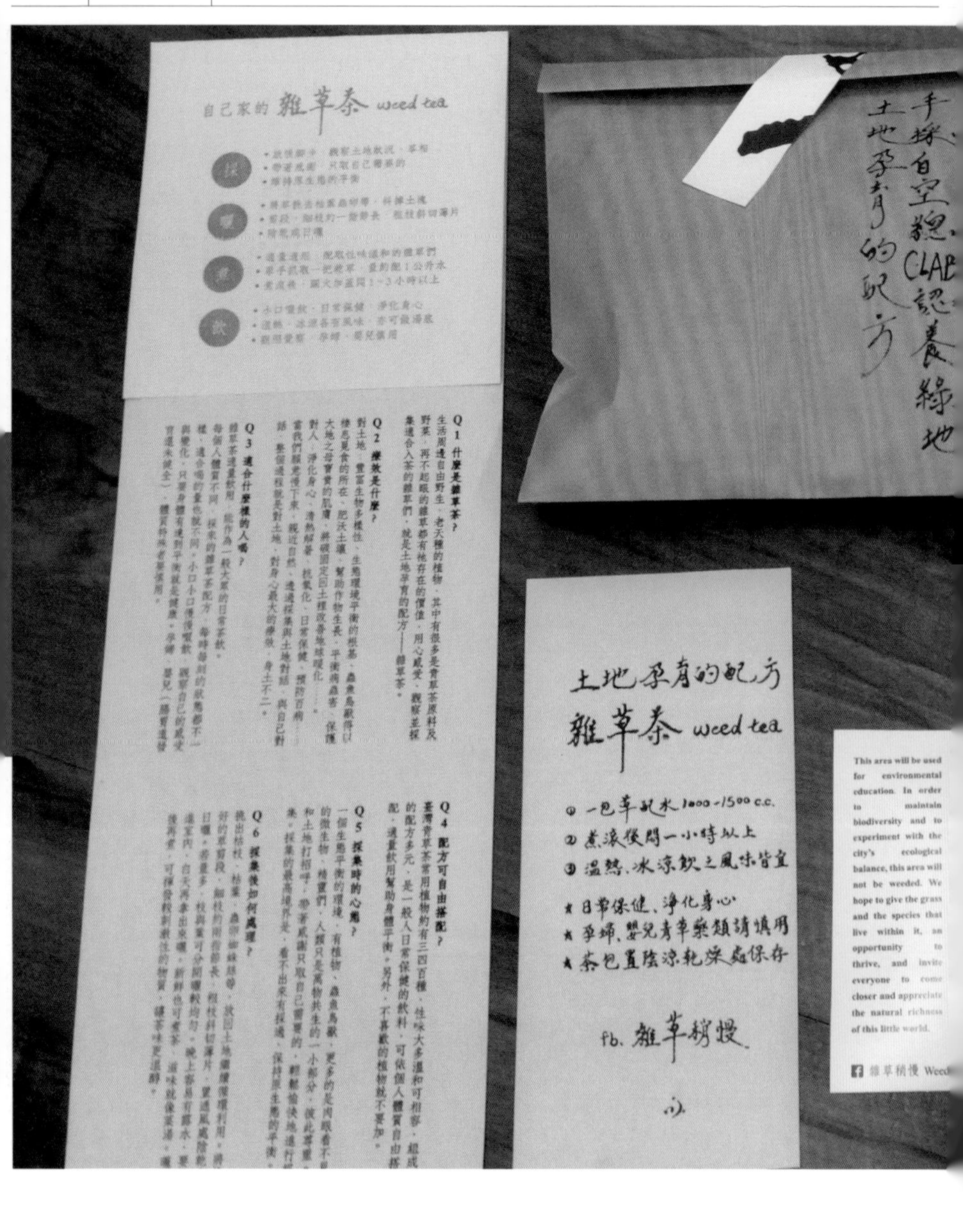
自己家的 雜草茶 weed tea
採
曬
煮
飲
Q1 什麼是雜草茶？
生活周遭自由野生、老天養的植物，其中有很多是青草茶原料及野菜，再不起眼的雜草都有祂存在的價值，用心感受、觀察並採集適合入茶的雜草們，就是土地孕育的配方——雜草茶。
Q2 療效是什麼？
Q3 適合什麼樣的人喝？
Q4 配方可自由搭配？
臺灣青草茶常用植物約有三四百種，性味大多溫和可相容，組成的配方多元，是一般人日常保健的飲料，可依個人體質自由搭配，適量飲用幫助身體平衡。另外，不喜歡的植物就不要加。
Q5 採集時的心態？
Q6 採集後如何處理？
手採自空總CLAB認養綠地
土地孕育的配方
土地孕育的配方
雜草茶 weed tea
① 一包草配水 1000-1500 c.c.
② 煮滾後悶一小時以上
③ 溫熱、冰涼飲之風味皆宜
★ 日常保健、淨化身心
★ 孕婦、嬰兒青草藥類請慎用
★ 茶包置陰涼乾燥處保存
fb. 雜草稍慢
This area will be used for environmental education. In order to maintain biodiversity and to experiment with the city's ecological balance, this area will not be weeded. We hope to give the grass and the species that live within it, an opportunity to thrive, and invite everyone to come closer and appreciate the natural richness of this little world.
雜草稍慢

▌雜草稍慢 / 林芝宇，「當代城市採集——自己家的雜草茶」，2019_

Weed Day/Zo LIN, *Urban Gatherer - Making Your Own Weed Tea*, 2019_

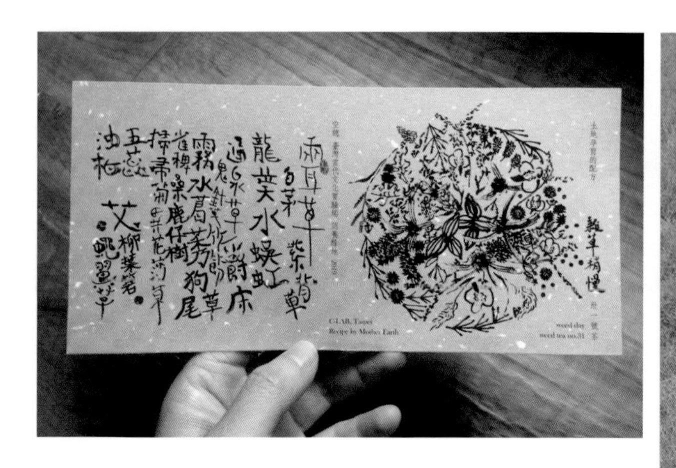

> 因為喜歡喝青草茶而跟著草藥師上山採草，因為看見了深山裡的除草劑而決定為土地煮一鍋茶，因為雜草們，開始發現人與土地的秘密……從白皙胖嫩的平凡上班族，到吃草吃土的採集藝術家， 療癒自己、療遇人心、療育土地。

> 大自然會自己長出適合這塊土地的植物，也許是不經意的風、也許是蟲魚鳥獸，也或許早已在土壤裡沉睡多年，這些細小堅韌的草籽，不大挑剔所在，物候風土合宜，就恣肆生長，人們不知其名，稱之「雜草」，猶記那是過去口耳相傳的青草藥或野菜，人與土地最單純的交往。適合入茶的雜草便是「土地孕育的配方」，採集日曬、熬煮成茶。就像小時候阿公阿嬤隨手從田裏抓草藥回家煮青草茶的味道，這些長在身邊、垂手可得的植物，多半清熱解暑去濕氣，適合處在氣候潮溼悶熱的臺灣人，土地最瞭解妳 / 你需要什麼，就是這樣熟悉的味道。

> 自由野生的雜草，孕育豐富的生物多樣性，時而靜謐，時而狂放，姿態如實。將祂們一個個挖掘出名字，重現人類自古以來與之相處的記憶。裡頭有許多青草茶原料、野菜、生活民具的材料，也有強勢入侵植物、有毒植物等等，許多雜草的價值甚至還未被發現，充滿無限可能。邀請你，腳步稍慢，陪伴自己與土地一起呼吸對話。從辨認、挑選、採集、剪段、晾曬，慢煮，感受生命的流動，觸摸大地與天空，品茗一杯在地風土的雜草茶。

> 當代採集是永續生活的契機，回歸當下，重新建立人與土地的心關係。這些土地自由野生之野花草木，有原生植物、青草藥、野菜，也有強勢入侵植物、有毒植物等；卻也具有生態系服務之功能，調節微氣候、淨化空氣，幫助水土保持，提供生物棲地孕育生物多樣性等。透過有意識的「野採」來維護自然的生態特性與環境棲地，復育環境生態，以無限的認識與觀察，有限度的採集，來創造人與自然環境的共好豐盛。

雜草稍慢 Weed Day 網站請見
https://www.weedday.org/

C-LAB 園藝——打造一個永續的系統
C-LAB Gardening — Growing Sustainable Systems

本計畫為「C-LAB 園藝——打造一個永續的系統」的研究計畫，期待透過研究、講座以及線上節目，逐步建立一個跨領域的社群，邀請包括藝術、建築、農業、設計、飲食與科技領域的專業工作者，展開互動與對話，試圖將從理念到行動、從個人到社群、從生產到消費等層面，探討「永續生活」可以被實踐的各種可能性。此計畫以 C-LAB 空間作為基地，打造一個線上實驗花園，透過園藝式的實踐，團隊期許能扮演「異花授粉」的角色，媒合跨領域的創意文化工作者，孵育可被落實的創意提案。

The project, C-LAB Gardening — Growing Sustainable Systems, supplemented by research, talks, and online programs, aims to introduce sustainability into all aspects of cultural practice, laying the groundwork for current and future questions of sustainable systems. The team plans to build a cross-disciplinary community of professionals from the arts, architecture, agriculture, cooking, design, and technology, to cultivate a dialogue and explore sustainable lifestyle. We will cover the spectrum from concept to action, from individual to community, and from manufacturing to consumption. The project hosts a virtual garden where all are welcome. This garden is expected to cross-pollinates ideas, bridge cultural divides, and incubate practical proposals from seeds to blossoms.

CREATORS

王維薇、蔣慧仙
WANG Wei-Wei, Sophie CHIANG

這是一個有機的組合，長期在文化和藝術領域耕耘的兩個朋友，王維薇和蔣慧仙想一起做點有意思的事情，再加上程式設計師 Christopher ADAMS，三個人分別從藝術、社會、科技的角度，共同來探討永續發展的議題。

It all started with an organic combination of two friends, WANG Wei-Wei and Sophie CHIANG, from the arts and culture fields, looking for a fresh project. Teaming up with Christopher Adams, we seek to explore sustainable systems from artistic, social, and technological points of view.

C9
王維薇、蔣慧仙、林暐翔、王韋智、Christopher ADAMS、Amie KO

王維薇、蔣慧仙，「C-LAB 園藝——打造一個永續的系統」，2019_
WANG Wei-Wei, Sophie CHIANG, *C-LAB Gardening – Growing Sustainable Systems*, 2019_

移動花園
MobileGarden

Welcome to our Garden!

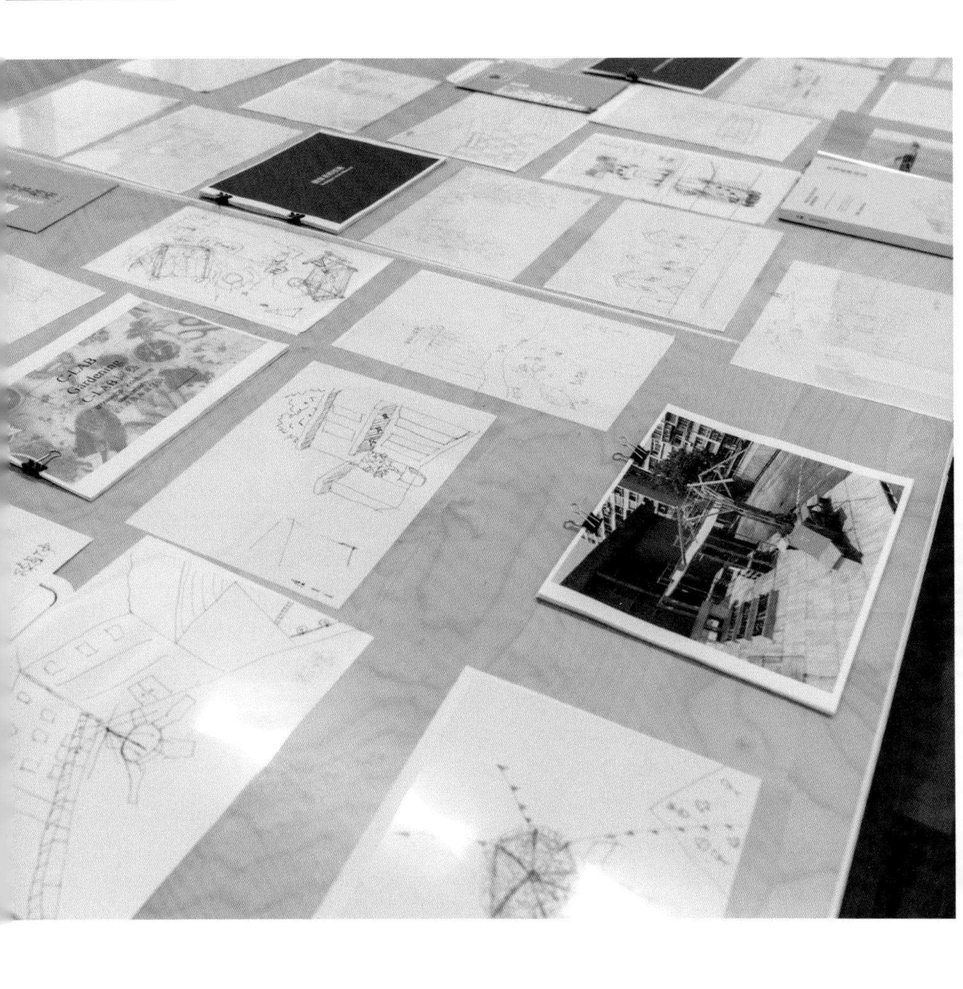

「C-LAB 園藝——打造一個永續的系統」計畫網頁請見
https://clab.org.tw/creators/2019-r6/

規訓的星系：戰後外省「離散」及「鎮壓」史調查創作計畫

The Galaxy of Discipline: A Survey of Post-war Diaspora and Suppression

本計畫藉由 1949 至 1987 年間政治檔案的分析，初步釐清戰後空總在戒嚴時期的空間運作性質，同時對陸、海、空軍的白色恐怖軌跡和差異建立基本認識。通過對今日稱為「外省人」的群體「去族群化」的過程，藉由密集的講座、讀書會、影片欣賞，使得戰後離散群體內部在階級、族裔、文化、政治和戰爭經驗的異質性得以突顯，進而能重新探究離散者認同的多線軌跡，並回答為何當年有大量外省人於戒嚴時期遭遇白色恐怖的歷史謎題。此外，通過訪談 1950 至 1970 年代曾在空總任職的軍法、政工人員，得以具體識別出當年軍法處、押房、法庭的確切位置，且通過口述進一步重建當時的日常運作概況。在研究分析與論述成果的基礎上，影像創作計畫將重新讓那些在「族群刻版化」想像中被遺忘、消失、殞落的「軍籍異端者」身影再次浮現於公眾眼前。藉此，亦能真正賦予「空總」及其他同類「規訓離散者」的遺址（sites）作為「異托邦」之地的能量。

By analyzing political archives between 1949 and 1987, this project attempts to clarify the operational nature of the former Air Force Command Headquarters during the Martial Law period and to construct basic understanding towards the military trajectory of land, air, and sea during White Terror, and clarify the differences within. By de-ethnicization through the concept of "Mainlanders," intensive lectures, book groups, and film viewing sessions enhance the disparities in class, racial, cultural, political, and warfare experiences and further explore the discrepancies between various self-identifications of the diaspora. The curation also addresses the historical mystery of why a large number of Mainlanders also fell victim during White Terror. In addition, by interviewing military law personnel and political workers who served at the Air Force Command Headquarters between 1950s and 1970s, the precise locations of the military law department, detention room, and courtroom were identified, allowing the reconstruction of daily operations through oral descriptions. Based on research, analysis and discourse, the video production project allows military heretics who have either passed away, disappeared, or been forgotten amidst stereotypical ethnic imaginations to reappear before the public. This also empowers the Air Force Command Headquarters and

other similar sites that were previously established to discipline diaspora to be transformed into "Heterotopias."

CREATORS

安魂工作隊 Libera Work-Gang

成立於2018年底，由一群創作、研究、及各行業勞動者組成。「安魂」具有雙重意涵。第一：安撫被遺忘的魂魄。我們想將消逝在白色歲月的鬥魂迎回故土，為故鄉的後輩看見。第二：重振當代人的靈魂。藉由認識往昔的抗爭，重探當代的景況，並且重新發掘與認識我們身上擁有的力量。為了找到這條路，所以各種位置的行動者彼此連結，最後形成了「安魂工作隊」。

The Libera Work-Gang was founded at the end of 2018 by a group of people working in the fields of art, academia, and various other professions. The term, "libera," holds two meanings. The first is to console forgotten souls. This echoes with the collective's objective to bring the spirits of White Terror fighters back to their homeland and for them to be recognized by their descendants. The second significance behind the term is to revive the spi-rits of contemporary people. Through learning about the resistance that took place in the past, present conditions are reexamined, as we rediscover and learn about the strength that we possess inside. In order to find this way, activists from various backgrounds have come together to connect with one another, which has led to the founding of the Libera Work-Gang.

▍安魂工作隊，「規訓的星系：戰後外省『離散』及『鎮壓』史調查創作計畫」，2019_

Libera Work-Gang, *The Galaxy of Discipline: A Survey of Post-war Diaspora and Suppression*, 2019_

> 以「安撫被遺忘的魂魄和重振當代人的靈魂」為宗旨而成立的「安魂工作隊」，是由來自不同領域背景的專業工作者自發組成的工作隊伍，以考察白色恐怖歷史為主要使命。計畫主持人林傳凱為白色恐怖歷史調查的專業學者，多年來致力於研究戰後初期的抗爭與國家暴力史，並訪談過三百餘位昔日的政治犯與受害者，同時亦對相關歷史的推廣教育活動不遺餘力。創作計畫主創者洪瑋伶長期在紀錄片與影像領域耕耘，製作過為數不少的影像企畫。夥同一樣是影像工作者的辛佩宜，與藝術家李佳泓，兩人亦都有著豐富的影像製作經驗。可見對歷史敘事的影像化是為團隊擅長並一再演練的關鍵核心實力。

> 「規訓的星系：戰後外省『離散』及『鎮壓』史調查創作計畫」裡頭暗藏著的是諸多繁複、難度甚高的挑戰。不僅是因為歷史研究本身就極其艱辛，還要為此做長時間的田調與勘查，並非一般人所能觸及並能就此有所理解的。白色恐怖中約有高達四成的政治犯為外省籍，其中又以軍籍身分為主，團隊在臺灣當代文化實驗場的創作計畫把視角聚焦於外省政治犯在時間上特殊的離散狀態，以及國家暴力對其鎮壓的未明空間敘事。

> 計畫期間，團隊主辦多場關於白色恐怖主題的讀書會，閱讀文本多為政治犯的回憶錄。其中主題如1950 年代由工人、醫護人員、師生、戲劇人士等推動的左翼政治新劇和監獄中的反共話劇；三的白色恐怖歷史圖像及其概論；現今國立臺灣大學水源校區的「國防醫學院舊址」對無人認領的治犯屍身處理；劉吉雄導演的紀錄片《寶島夜船》，內容為報導陳松如何從「反共義士」變成「治犯」的過程；白色恐怖下的「僑界」肅清與外一種的外籍政治犯；美軍駐臺時的戰爭與色情產業還有海軍政治犯馮馮潛藏的情慾身體史；作家胡淑雯分享白色恐怖題材相關的文學小說與史料；總紀念政治犯的第一場活動——江國慶的枉死與「測謊機」作為審訊技術的實效性；最後一場推活動則找來表演藝術工作者，帶領學員進行名為「找回感覺」的身體工作坊，旨在體驗被規訓的體記憶，以及在生活空間中無形或有形的受壓迫感。（文｜謝鎮逸）

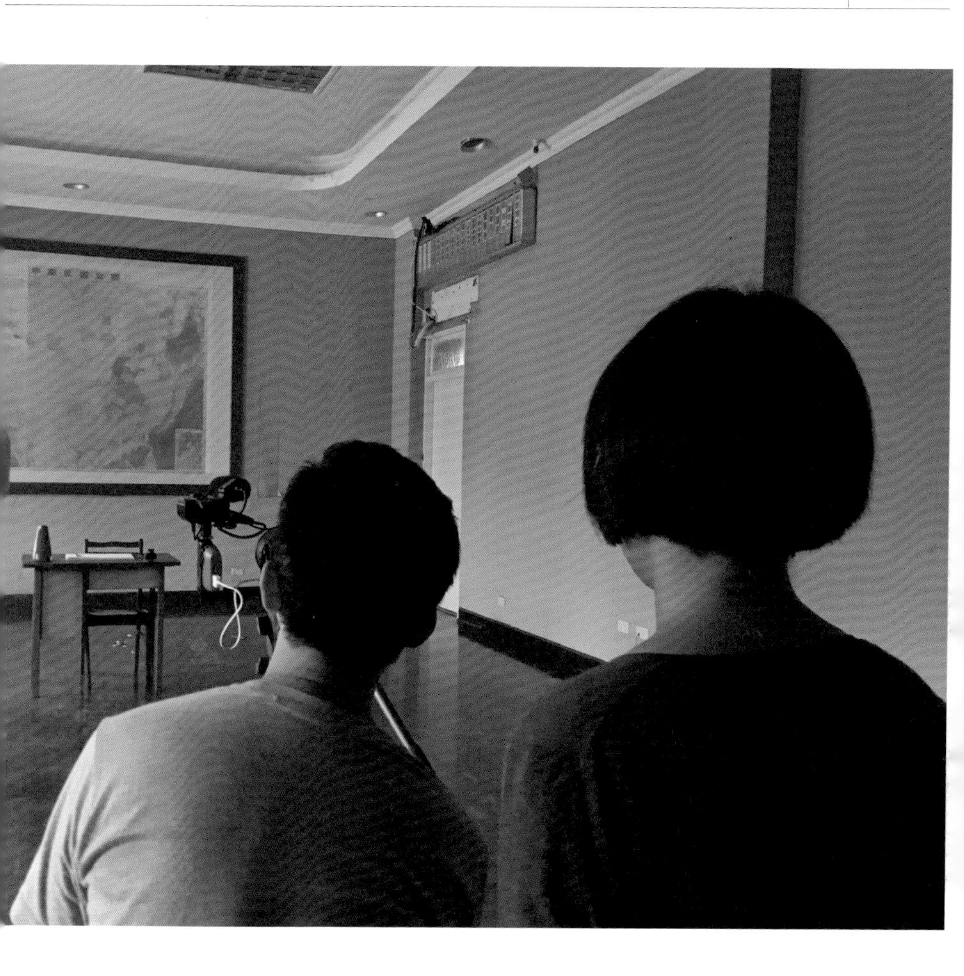

觀察報告

謝鎮逸〈過去從未過去，「安魂工作隊」的調查行動與歷史時空再考掘〉，全文請見
https://mag.clab.org.tw/clabo-article/libera-work-gang_the-galaxy-of-disciplinelibera-work-gang_the-galaxy-of-discipline/

再現・抵抗・瓦解：一次重訪臺灣同志污名史的邀請

Representation, Resistance, Reconciliation: Tracing the History of Tongzhi Stigma in Taiwan

本計畫首先欲回到1990年代，回溯同志運動剛開始的時候，除了曾在歷史上留名的知識分子與推動社會運動的同志之外，其他的同志又在哪裡？過著什麼樣的生活？他們與檯面上的運動軌跡及公眾空間發生過什麼關係？本計畫欲挖掘這些相對私領域、私密記憶的同志文化，納入現有、已被書寫記錄的同志文化歷史地景，呈現更為複雜且多樣的污名地圖。

此外，本計畫欲邀請各領域的非同志創作者，先共同走踏、聆聽同志空間歷史，而後翻轉同志故事的訴說——接收、書寫——受眾的上對下關係，轉由不同專業的創作者設計互動工作坊，嘗試尋找不同觀點下的污名空間敘事，推展共同創作的各種可能性。我們希望以此計畫作為社群內外扣聯、彼此滲透的途徑，打開同志以身分認同為框架的族群記憶，得出更多樣且有機的生態切片。進一步詢問：同志運動行走迄今，我們還能如何以新的角度和大眾勾連？

This project first returns to the 1990s to trace the beginning of the Tongzhi (LGBT) Movement. Apart from intellectuals whose names have been recorded in history and other Tongzhi members who were driving forces of the social movement, where are other members of the Tongzhi community? What sort of lives do they lead? What are their relationships and experiences with the visible trajectory of the movement and the public space? This project attempts to dig into the domains and memories of Tongzhi culture that are relatively intimate and to insert them into the landscape of written records of Tongzhi history, presenting a more complex and multivariate map of stigmatization.

In addition, this project plans to invite non-Tongzhi (heterosexual) creatives from different fields of expertise to trace and listen to Tongzhi space and history together, finally reversing the "declaration – reception/writing – audience" hierarchy of Tongzhi narrative. This project presents an interactive workshop designed by professionals of different fields and searches for stigmatized space narratives

that originate from varying perspectives, developing the possibility of collective creativity. We hope that this project will become a channel for connecting and immersing different communities, opening a collective memory of Tongzhi identity and producing more organic and multifaceted understandings. Furthermore, this project also explores ways and angles to connect with the masses at current stage of the Tongzhi movement.

REATORS

時文化有限公司 Cultime.co

時文化有限公司成立於 2018 年，由一
當代藝術、文學、電影、設計等不同領
的藝文工作者所組成的工作團隊。我們
由創造性的途徑，看見公眾視野之外的
灣歷史，透過文化生產與人員培力，轉
知識為可共享之資源。

Established in 2018, Cultime.co is a working team comprised of workers from artistic fields includes contemporary art, literature, cinema, and design. We rediscover Taiwanese history peripheral to the public through creative methods. By means of cultural production and professional empowerment, we transfer knowledge into shared resources.

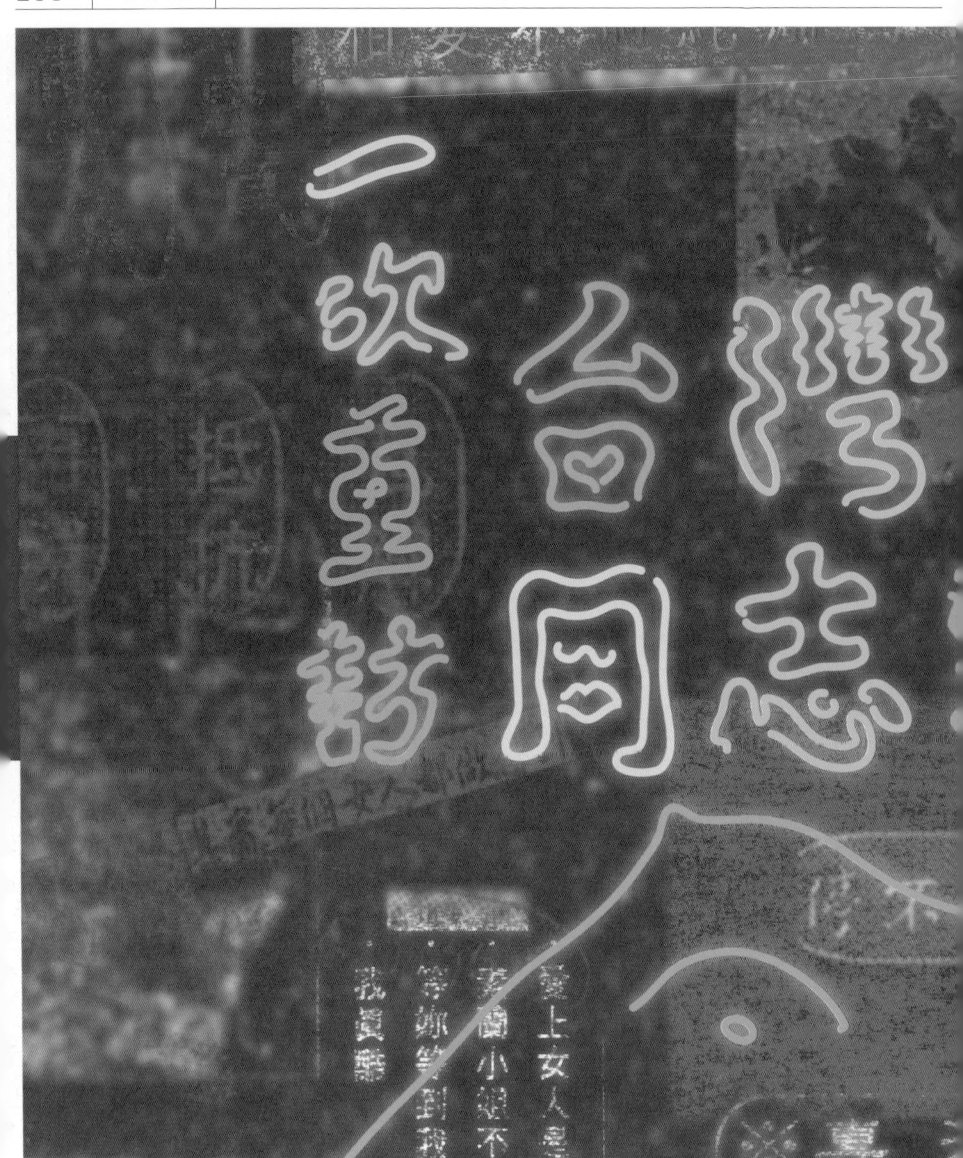

▌沃時文化有限公司，「再現．抵抗．瓦解：一次重訪臺灣同志污名史的邀請」，2019_

Cultime.co, *Representation, Resistance, Reconciliation: Tracing the History of Tongzhi Stigma in Taiwan*, 2019_

› 沃時文化有限公司所提出的「再現・抵抗・瓦解：一次重訪臺灣同志污名史的邀請計畫」，意在進行文史調研及具特殊意義的資料庫構建。從臺灣同志空間史出發，連結相關文獻及大眾活動，反應族群內群集的認同力量。

› 2020 年 6 月 12、13 日沃時文化有限公司舉辦為期兩日的「發掘女同志關鍵字：史料讀寫工作坊」，由蔡雨辰、陳佩甄、陳韋臻、吳淳畇共同帶領。沃時文化有限公司長期關注性別少數議題，至 2019 年工作團隊盤點臺灣有關多元性別的歷史資源盤點與口述記憶採集，發現在國家級歷史博物館或是檔案館中同志相關史料文物依舊缺席，顯示該群體的相關記憶仍未進入臺灣歷史集體記憶的書寫範疇。該計畫嘗試將女同志相關史料與文物置於公共史學的框架下，以女同志為主軸進行記憶蒐集與共筆書寫。

› 工作坊操作方式是將近年所徵得的女同志相關私人蒐藏作為素材，提供給來自臺灣各地不同世代、不同領域及不同性傾向的學員共讀與共筆，藉由對於史料與文物的認識與描述，重新連結過去斷裂或幾近消失的歷史片段。活動進行依循計畫中的四大主題：集結、現身、空間、婚權將學員分組，從瞭解如何撰寫詮釋資料、鎖定物件與議題關聯以及關鍵字的設定等步驟導入，各組依組別主題進行共筆撰寫。其中不少取材自《女朋友》雙月刊，該刊物是由目前所知最早的女同志組織「我們之間」製作，發行期間透過郵寄訂閱方式深入全臺鄉間、漁村，是當時各地女同志映照自身認同、瞭解運動狀況及交友聯誼的重要管道。除該套刊物之外，其它散見的文件、書籍及物件，學員普遍對於這些留存的內容感到意外，且認為親身接觸這些珍貴文物是理解 1980 至 1990 年代女同志運動的重要關鍵。

「再現・抵抗・瓦解：一次重訪臺灣同志污名史的邀請」臉書專頁
https://reurl.cc/9GVanV

「女同志史料蒐集撰寫計畫」國家文化記憶庫資料網頁
https://reurl.cc/GxeOGZ

磁帶音樂記譜及模組研發計畫

The Research Project of Scores for Tape Music and H.D.C.M. (Human Dynamic Co-ordination Models)

「磁帶音樂記譜及模組研發計畫」為《人體同步模組》計畫系列之一。《人體同步模組》的工作，從技術角度來看，便是將傳統科技哲學「機械係人體延伸」觀念反轉，將媒體機器運作，還原為人體動作與聲音。例如將泡泡演算法、細胞自動機、磁帶機、音樂盒之數學及機械結構，轉化為人與人之間的互動協定，我們將此方法稱為「模組」，即「原型」之意。

「模組」緣起於跨媒體藝術家林其蔚 2004 年之《磁帶音樂》演出，自 2017 年開始，林其蔚構思發展磁帶音樂以外的身體互動模式，並逐步發展成為團體創作計劃，邀請電影導演陳芯宜等創作者，參與全新模組的研發與測試排練。「模組」意圖創造一種特殊的團體性，在共同遵守一組規則的前提下，為了創造一個演出，展開內部的激盪。參與者在此間，可以從眾，可以獨創，這樣的表演形態亦可邀請觀眾為參與者，讓來自不同背景的參與者，開發各自的欲望，同時又能夠在互動間呈現出集體性，將這兩者的辯證，呈現在最後的聲音與肢體美學中。

The Research Project of Scores for Tape Music and H.D.C.M. is part of the Human Dynamic Co-ordination Models. From a technical perspective, Human Dynamic Co-ordination Models is a reversal of the concept that "the machine is the extension of the human body" in traditional philosophy of technology, reversing the operation of media machines into human movements and sounds. For instance, the mathematical and mechanical structure of the bubble sort in algorithm, cellular automation, cassette machine, and music boxes are transformed into agreements for human interaction. We refer to this method as "models," which means "prototype."

The term "model" originates from intermedia artist LIN Chi-Wei's *Tape Music* performance in 2004. Since 2017, LIN Chi-Wei developed alternative bodily interactive methods distinctive from tape music, which gradually grew into a group creative project. The project invites creatives including film director Singing CHEN to participate in a whole new development and testing rehearsal. The "model" intends to create a unique community which forms a performance that embarks on inner exploration on the premise of following a set of rules. Participants can

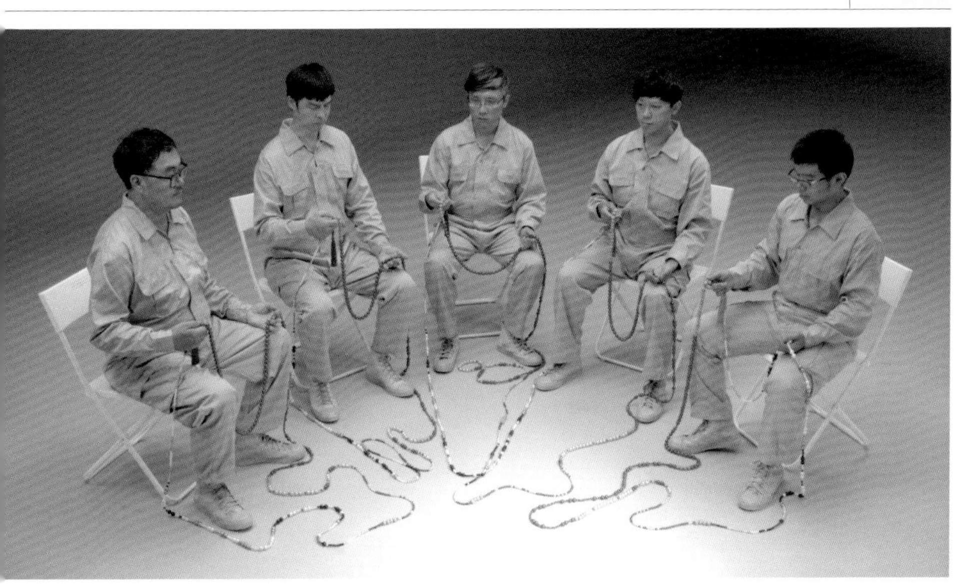

follow the crowd or forge their own path, while the audience can also be invited to take part in the performance. This method allows participants from different backgrounds to cultivate their individual desires while forming collectiveness throughout the interactions. Finally, the dialectical interactions between the two are displayed in the final sound and bodily aesthetics.

CREATORS

林其蔚 LIN Chi-Wei

1971 年出生於臺北，曾研讀法國文學、臺灣傳統藝術以及新媒體藝術創作，參與噪音實驗團體「零與聲音解放組織」，並於 1990 年代擔任各種地下噪音節目策劃。於研究所時期，他廣泛地參與了民間儀式音樂、廟宇雕刻的田野。2004 年以降，林其蔚開始了一系列他稱之為「聲音模組」的跨媒體實驗，從人類傳播技術史尋找靈感，將結繩紀事、符號、圖象、文字與類比、數位技術統合，以人體為機械元件，在沒有導演，沒有指揮的前提下，組合成集體聲音機器，同步運作並構成音場。

Born in Taipei, 1971, LIN Chi-Wei is a transdisciplinary artist who has received academic training in French literature, cultural anthropology, and media art. He participated the noise band Z.S.L.O. and was responsible for the programming of various alternative art festivals during 1990s. LIN also explored the realms of religious music and temple sculpture through field researches when he was studying for his degree. Lin initiated a series of transmedia experiences which he names “sound module” from 2004. He drew inspiration from the history of technological communication and combined knotted strings, symbols, images, languages with analog and digital technology. Meanwhile, human body is also transformed into a piece of mechanical element and integrated into a collective sound device, thus constructs a sound stage without a composer.

林其蔚，「磁帶音樂記譜及模組研發計畫」，2019_

LIN Chi-Wei, *The Research Project of Scores for Tape Music and H.D.C.M. (Human Dynamic Co-ordination Models)*, 2019_

> 「磁帶音樂」（又名「音腸」）系列，邀請多人圍坐成螺旋狀，傳遞著長長的帶子並念誦著上頭由不同中文單字搭配英文拼音的記譜，進而在這臨時組成的群體中形成某種音團同步的一致性。觀眾是參與者同時也是表演者，因現場當下的同步發音互動來創造出有機的曲子。在此時此地的空間中，藉由集體跟隨著指令發展出來的規律，身體感將逐步成就某種儀式性，但之中偶爾也會有隨機性的節奏破壞和沉默滲入而隨時產生變動。這不僅是純粹聲音測量與作曲手段，卻又同時指涉群體互動特性的社會意涵。

> 林其蔚最早以這種模組形式出現的作品雛形源自於運用約翰．何頓．康威（John Horton Conway）的同名數學原理來運作的一台音樂機器《細胞自動機》（cellular automaton）。在歷經多次的發展與版本更新，「磁帶音樂」開始轉向成為一套可供持續開發的「模組」。於是，「人體同步模組」的概念因此在日後被逐漸確立，並在 2017 年正式成為一系列的創作計畫。

> 使用人體作為取代機器的模組，再採取務求簡單的互動規則，來進行團體運動；而該計畫目前的主要實踐場域是表演藝術。在他的設想中，這組互動規則能夠取代今日表演藝術實踐中，音樂之樂譜、舞蹈之舞譜、戲劇之腳本等功能；甚至可以取代時鐘，音樂數位介面（Musical Instrument Digital Interface）訊號，配樂和燈光在現代劇場中的同步功能。而應用之策略及方法，是在每一個團員遵守一組互動規則的前提下，大家一起運作這個程式；而團員可以在現場集體決定演出的速度和節奏，甚至即興提供演出的內容。換言之，作為運作邏輯腳本「譜」的運作流程和規則指令上是給定的，但卻能夠因不同的演奏者和情境設定而生產出不同的範式及其結果。進退與差異的變化中是允許規則的可逆、可反覆性，以及非線性的發展。而不同的參數與數列加入，都會導致產生不同的多次方變化。（文｜謝鎮逸）

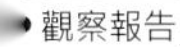
觀察報告

謝鎮逸〈超越現代性分工邏輯的身體機器，從「磁帶音樂」到《人體同步模組》〉，全文請見
https://mag.clab.org.tw/clabo-article/human-dynamic-co-ordination-models/

延遲青春
Delayed Youth

「延遲青春」將情境設立於 2050 年的臺灣，在這個未來世界裡，保守團體透過公投以及積極參政，成功地掌握政治資源，並策動衛福部去改變臺灣關於性別的政策制定。其中他們透過與新陳代謝科和小兒科醫師的合作，將治療性早熟的藥物應用於青少年的性成熟控制，政府與地區醫院合作，定期定量的發放此藥劑給家長，讓家長固定於家中施打於自己尚未開始青春期的小孩身上。

透過醫療科技的應用，政府可以透過此藥物，讓全國所有的青少年延緩其青春期的發生。所有的小孩將會被政府統一控制於 18 歲開始青春期，也就是說透過這樣的過程，可以有效地讓身體的性成熟時間與法律上合法性交的年齡同步。這也意味著在高中以前的所有青少年，都不具有任何第二性徵，也不會因為荷爾蒙的作用而產生求學期間不需要的性慾，藉此降低所有因為「性」所產生的問題。此計畫透過一系列反烏托邦的未來情境想像，去建構一個極端的虛構世界，並以這個被建構的世界作為一個溝通的平台，促使社會大眾對於青少年的主體以及「性」的議題，進行更深度的思考。

Delayed Youth is set in the year 2050, Taiwan. In this futuristic world, conservative groups have successfully taken hold of political resources through referendums and by taking an aggressive stand in political participation, and are instigating the Ministry of Health and Welfare to change policies on gender and sexuality. By collaborating with the Endocrinology Department and pediatricians, these groups use medication for sexual precocity to control sexual maturity among adolescents. The government works alongside local hospitals to distribute medicine on a regular basis so that parents can give the medication to children who have not yet reached puberty.

By applying medical technology, the government is able to postpone the puberty of all adolescents in the nation. Children only reach sexual maturity at the age of eighteen; in other words, this treatment allows the stages of sexual maturity to synchronize with the lawful age of sexual intercourse. This means that no adolescent will have secondary sexual characteristics before reaching high school, reducing sexual desire brought forth by hormones during the student years, therefore lowering the likelihood of sex-related problems. This project presents a series of dystopian scenarios to construct a fabricated

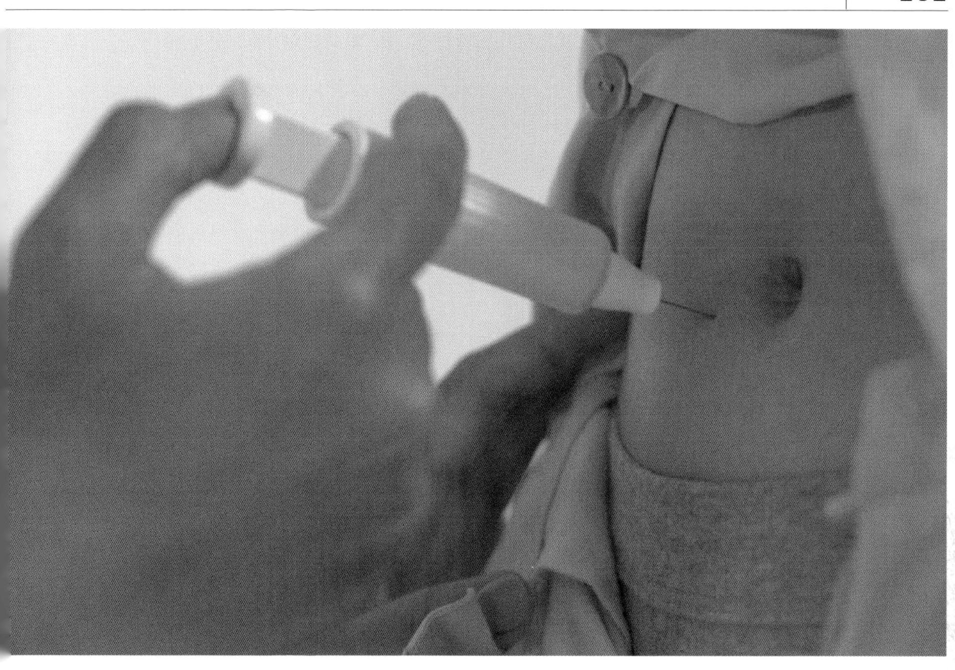

world of extremes, inspiring the masses to contemplate on the individuality of adolescents and the issue of sex.

REATORS

廣毅 KU Kuang-Yi

生於臺灣臺北，碩士畢業於荷蘭恩荷設計學院社會設計研究所、國立陽明學臨床牙醫學研究所、實踐大學媒體達設計研究所，具有牙醫師、生物藝家以及推測設計師等多重身分。他試拓展藝術、設計與科學結合的可能性，品主要專注於臨床醫學、人類身體、與其他物種的關係以及性別議題，嘗藉由藝術實踐與設計方法去探索科學域中的倫理問題，並藉此思考科技、類個體和環境之間的關係。

Born and raised in Taipei, Taiwan. Ku graduated with triple master degrees in social design from Design Academy Eindhoven; in dentistry from National Yang-Ming University; and in Communication Design from Shih Chien University. Formerly a dentist, he is a bio-artist and speculative designer. He tries to explore the possibility to combine art, design, and science. His works mainly deal with clinical medicine, human body, the relationships between human being and other species as well as gender issues. He manages to tackle ethical problems embedded in scientific fields by means of artistic practice and designing, thus to probe into the relationships among technology, individuals, and the environment.

顧廣毅，「延遲青春」，2019_
KU Kuang-Yi, *Delayed Youth*, 2019_

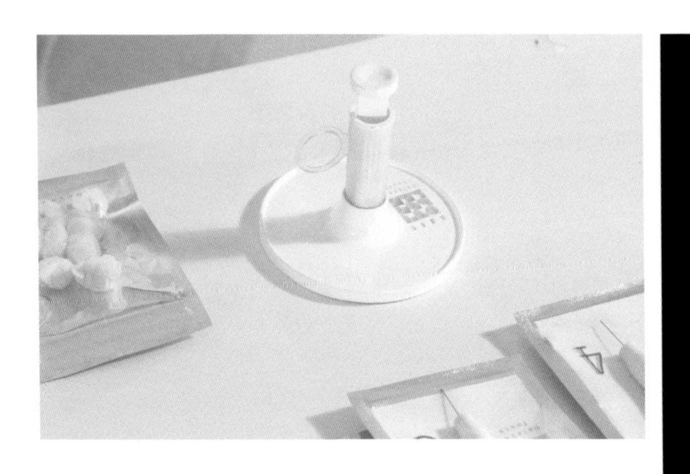

> 「延遲青春」中的虛構設定，來自當今臺灣社會持續發生關於性平與教育政策的對立觀點與爭議，這些分佈在光譜兩端的激辯多半來自於性別保守團體與性別平等教育團體雙方針對「人類對『性』的掌控程度」的認知態度與歧見，亦關乎身體性成熟的實際情況以及現行法律對於性行為的發生，兩者在年齡規範上的矛盾。

> 「延遲青春」著墨於情境結果的直接展示，包括一支擬仿政令宣導的「延遲青春專案」衛教影片展示官方的解說、藥物施打的示範與視覺象徵的畫面，同時設計出該政策下的制服、說明手冊與針具裝置的細節。為貼近如長照、戒菸或反毒議題般的政令宣導影片質感，顧廣毅與平面設計師合作，為「延遲青春專案」設計出一套專屬於該政策的獨立視覺系統。這些視覺元素，也在計畫期間成為臺灣當代文化實驗場的「城市震盪」展覽作品之一，走進原先是軍用辦公空間的戰情大樓展出現場，內側以及靠近門邊的牆面上仍分別保留著一幅中國地圖與世界地圖，在幽暗的燈光下，「無性形象」制服的懸掛狀態頗有博物館展示特殊朝代服飾時，為將細節明顯呈現的手法。

> 雖然是針對未來社會情境的想像，但計畫裡所使用的注射藥物卻是現在就已作為規範使用中的正式醫療藥品，透過這樣的計畫凸顯假設情境中弔詭的政策與回應社會氣象。比起著重尚未發生的科技所帶有的科幻感（science fiction），顧廣毅提到這樣的創作概念更偏向「推想小說」（speculati fiction）或者同樣傾向探索未來情境且不以解決問題為最終導向的「思辨設計」（Speculati Design，或譯推測設計）範疇。配合工作坊中醫療專業知識的導入，某種程度上亦加深了此未來情境的「可信度」與「可行性」。當性平教育在當前臺灣社會不時淪為政治操作的手段之一時，這樣的情境更為「延遲青春」增添一份諷刺預言的意味。然而，現實事件的荒謬往往不亞於情境想像的設定
（文｜黃鈴珺）

讓家長每月定期定量給孩子施打藥物

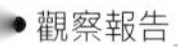
觀察報告

黃鈴珺〈許你一個無性的未來？顧廣毅《延遲青春》〉，全文請見
https://mag.clab.org.tw/clabo-article/delayed-youth/

無用便利店
s.s.s. Mart story/solution/smart

「無用便利店」是一個民眾參與的計畫，是一間不一樣的線上店。我們的「產品」出自你我再熟悉不過的小人物之手：可能是樓下早餐店的老闆娘、隔壁水果攤的小陳、菜市場裡的阿美姐等。每位設計者都各具特色，有的邏輯理性，精準掌握尺寸、細節，有的創意無止盡，以各種神展開的方式延伸、創造，但相同的是，他們都善用生活裡便利取得的材料，加以改造，製作出能解決生活大小問題的「產品」。

透過這個計畫，我們希望邀請大家仔細看看這些因著「生活」迸發、相當有意思的創意。在我們看來，這樣的「產品」富含文化底蘊的生存智慧，是難能可貴的生命積累，也是「無用便利店」希望珍惜與傳遞的價值。

s. s. s. Mart is a participatory project for the masses and a unique online store. Our "products" are made by familiar and ordinary people, such as the boss of the breakfast store downstairs, Mr. Chen from the fruit stand next door, or Mrs. Mei from the local market. Each designer has their own character; some excel in logical reasoning and are capable of grasping precise dimensions and details, while others have infinite creativity and can extend and utilize different methods. However, all designers share the ability to transform everyday materials into "products" that can tackle daily tasks.

Through this project, we intend to invite everyone to inspect these creative and interesting ideas inspired by life. For us, these "products" incorporate the living wisdom of our culture, which is precisely the value and concept s. s. s. Mart intends to deliver.

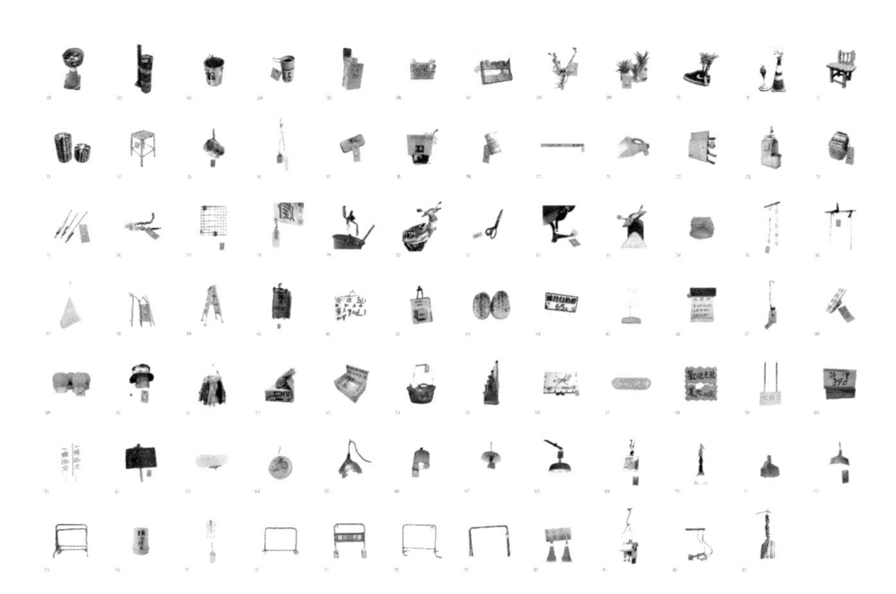

CREATORS

慢慢說 mamaisun

「mamaisun 慢慢說」團隊，以產品開發與藝術創作為基底，探索生活意義為指南，相信用心觀察，仔細聆聽，盡情想像，是生命最有用而無價的滋養。2017年成立迄今，不斷思考各種創新、有趣的方法，努力開展能 讓生活「有那麼一點不同」的好事。「mamaisun 慢慢說」希望透過充滿溫度的計畫與行動，創造人與人、人與物間，更良善的連結，為你我的生活 帶來多一點樂趣與溫暖，即便無用， 也能讓生活別有滋味。

Mamaisun is a team seeking to discover the meaning of life with product development and artistic creation. They believe that observations with a thoughtful heart, listening with an openmind, and imagining with all your passion are most useful and priceless nutrition of life. Founded in 2017, mamaisun keeps brainstorming for all kinds of innovation and fun methods to develop the good things that can make our lives a little different. They hope that, through a project and campaign of warmth such as mamaisun, they will create better connections between people, and even between people and objects, which will bring more fun and warmth to our lives. Even if something is good for nothing, it can make our life taste sweet.

▌mamaisun 慢慢說，「無用便利店」，2019_

mamaisun, *s.s.s. Mart story/solution/smart*, 2019_

新東松肉號
2391-2259 0918-365698
本館7號
用專鍋火
豬肉片 牛肉片 羊肉片
肉羊

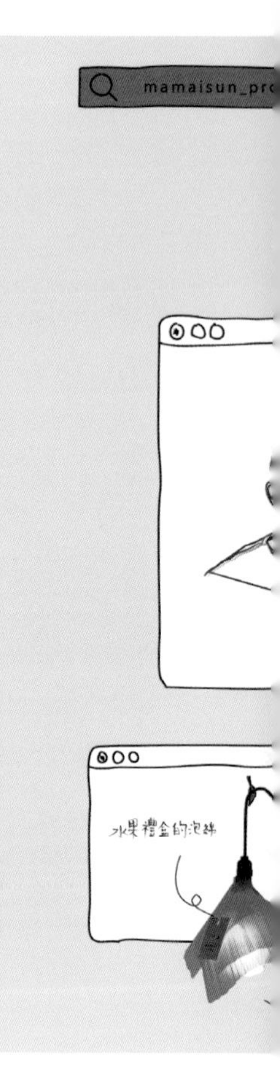
水果禮盒的泡棉

無用便利店
https://www.instagram.com/s.s.s.mart/

發酵城市
Ferment the City!

本計畫試圖從自然界的發酵原理及作用過程，重新埋解社會發酵的運作原型。藉由「發酵」作用，將廢棄物轉化為可用的資源與資材，探尋不同利害關係的人群可扮演之角色，及與其循環再利用模式。計畫成員在市政府提供的閒置空間「河神的丸子」彼此相遇，每一位都是獨立的個體又是因興趣連結而結合的群體。藉由協作與分工，圍繞著自然與社會的「發酵作用」展開研究與實驗。從農業科學、環境研究、都市規劃、社區到心理學的跨領域專業合作，激盪出「發酵作用」在有機物質、無機物質以及社會關係，如何可能串聯起來並展開循環、相互滋長的作用力。歷經六個月的工作坊及講座，以及許多實踐與行動，串起了網絡包括綠點點點點、不垃圾場、Beher 食物圖書館、穀盛有限公司以及許多市民朋友等參與，其中，與不垃圾場、綠點點點點共創塑膠布圓拱裝置也受邀參展「城市震盪：循環」展覽，現場將發酵城市的「知識圖譜」(knowledge mapping) 分享傳遞給更多民眾。

This project attempts to reinterpret the operation model of social fermentation through the principle and process of natural fermentation. Using the principle of fermentation, wastes are transformed into usable resources and materials. In addition, the project explores the role of human gatherings of different stakes and the method of cycling and reuse. Project members meet at Hó-Sńg House, an idle space provided by the city government. Each member is an independent individual who joined each other due to similar avocation. Through cooperation and division of labor, research and experiments of the “fermentation effect” that circle around nature and society are conducted. Through professional, interdisciplinary collaborations between agricultural science, environmental research, urban planning, and community psychology, this project explores how organic and inorganic objects as well as social relationships “ferment” and forge connections and complement each other. After six months of workshops, lectures, and multiple practices and operations, the project has attracted the participation of Our Green Map, Trasholove, Beher Food Library, Kokumori Foods, and many citizens. Particularly, the plastic cloth dome installation created with Trasholove and Our Green Map was invited to be a part of the City Flip-Flop: Circulation exhibition, sharing the knowledge mapping of fermentation city to more people.

REATORS

發酵城市 Ferment the City

計畫成員施佩吟 、陶曉航 、蘇映晨、énaïc PARDON、Claire MAUQUIÉ、tefan SIMON與 Lauren KIM一群人在社群基地「河神的丸子」彼此相遇，因為分享著相近的環境觀念及社會理念，觸發了團隊組成的起點。藉由不同專長和興趣的協作分工，圍繞著自然與社會「發酵作用」的可能性展開研究與實驗。計畫成員包括農業及環境學家、環境設計師、心理學研究者都市研究及社區營造工作者，彼此探索如何藉由發酵作用，將廢棄物轉化為可用的資源與資材。城市中有機廢棄物如塑膠製品、電器等，以及有機廢棄物的循環再利用，如何在社會過程的作用中，串聯起來並展開循環及應用組合。

Project members Peiyin SHIH, TAO Hsiao-hang, SU Yingchen, Lénaïc PARDON, Claire MAUQUIÉ, Stefan SIMON, and Lauren KIM met each other at a community base “Hó-Sńg house.” They founded the team because they shared similar environmental beliefs and social ideals. Through different expertise and interests, they cooperate, carry out research, and experiment on the possibilities of natural and social “fermentation.” The project members include agricultural and environmental scholars, environment designer, psychology researcher, and urban studies and community building professionals; they explore how to transform wastes into usable resources and materials through fermentation. How can organic wastes of cities, such as plastics and appliances, and the recycling and reutilization of organic wastes be connected to give rise to new cycles and application portfolio through the effect of social process.

施佩吟、蘇映塵，「發酵城市」，2019_
SHIH Pei-Yin, SU Ying-Chen, *Ferment the City!*, 2019_

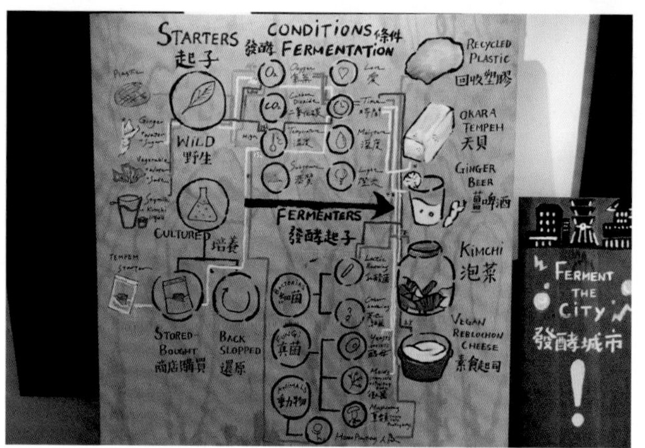

> 發酵需要環境和中介物質的作用，改變才會發生。改變一方面是物質轉化產生的質變，增添風味或是營養元素。因為空氣和大自然中本身就存在著酵母菌，在小瓶罐中的中介者有了啟動器酵母菌，給予環境後改變自然地發生。跳脫出小瓶罐的世界，回到了有人類存在的社會世界，我們以為改變了物質事實上因為著發酵期間帶來的等待，學習、交流、對話、磨合、熟悉、信任、默契……等，各種為人際增溫的互動產生，我們用等待創造了時間感，關係變得再也不同。

發酵城市 Ferment the City! 臉書專頁
https://www.facebook.com/fermentthecity

發酵城市 Ferment the City! 相關影片
https://www.youtube.com/channel/UCFABztPVnvl2CB23zuhZrTw/videos

社會運動交織而成的共同創作
Collectives Weaved with Social Movements

2018 年 9 月，苑裡市場被一場無名火燒毀，而後「畸零地」工作室與「掀海風」團隊在有著相同理念，運用不同的執行方式，開啟了與古蹟及文資保存的意識對話：畸零地工作室透過古蹟燒攤車計畫，走遍不同具有爭議性的古蹟建築，以販賣古蹟造型的雞蛋糕開啟與民眾的交流，進而民眾意識到文資保存相關議題。

而苑裡在地青年團隊掀海風從在地抗爭之後，從「反」到「返」，開始實踐青年返鄉運動，一起耕耘地方共生共創的社群。透過本計畫串連藝術工作者，藉由苑裡市場的歷史背景、審議式工作坊、公聽會，生產出參與式演出《苑裡好人》，探討好人的定義為何，又該由誰決定，反應市場的未來及地方城鄉所面臨之困境。

Yuanli Market was destroyed by a mysterious fire in September, 2018. With the same goal in mind, Groundzero and Hi Home used different methods to initiate dialogues between historical sites and cultural preservation. Traveling to various controversial heritage architectures with the Burn Down Cake Project, Groundzero sells sponge cakes shaped like the architectures to encourage dialogue with locals, hoping to inspire contemplation towards cultural preservation.

After focusing on resistance, local Yuanli youth group Hi Home took a detour from “resistance” to “return,” returning to their homeland and contributing their skills to locals. This project gathers art professionals, while the historical background of Yuanli Market, deliberative workshops, and public hearings lead to participatory performance *The Good Man of Yuanli*. The work explores the definition of a “good man” and who the decision-makers are, reflecting the future of the market and the challenges facing urban and rural areas.

REATORS

零地創造股份有限公司
roundzero

崎零地」意指我們希望扭轉某些事物在代社會的價值成見，用更開闊、永續的計思維以實現更好的普世價值。畸零地造集結了一群藝術與設計背景的當代新思想家與實踐者，共同建構出一個全方的創意工作團隊，擁有專業的議題挖、資料研究能力。藉由不同領域的跨界作，以匯集創作能量，勇於挑戰前人不嘗試的創作類型和議題。我們擁有自己造的鐵木工廠與設計工作室結合的工作間，從作品中處處能看見設計、工藝與術的痕跡。源自於我們堅持在自己的工親身動手，感受實體製作。每個細節貫著自身對專業的執著、對創作的熱情與實的執行能力。

"Zero ground" literally means the intention to reverse the value stereotypes of certain things in contemporary society, and use more open and sustainable design thinking to realize better universal values. Groundzero is comprised of a group of contemporary emerging thinkers and practitioners with backgrounds in art and design. Groundzero builds a comprehensive creative work team with professional issue mining and data research abilities. Through cross-border cooperation with different fields to gather creative energy, Groundzero dares to challenge creative types and issues that have never been tried before. Groundzero owns a working space combining our ironwood factory and design studio, and you can see traces of design, craftsmanship and art from our works. Thanks to Groundzero's persistence to practice in our own workshops to experience physical production. Every detail carries out its dedication to professionalism, passion for creation, and solid execution ability.

JANSPORT

▌畸零地創造股份有限公司，「社會運動交織而成的共同創作」，2019_

Groundzero, *Collectives Weaved with Social Movements*, 2019_

> 2017 年春天，位處竹圍的一間破敗成衣廠，經由設計與藝術創作者所組成的團體，一磚一瓦打造出一間可容納各種工事的共同工作室，以建築法中的術語「畸零地」為名，從樓下的木工、鐵工，到樓上的文書、設計、裁縫、攝影的複合式空間，各種創意和發想在此迸發。同年的臺北白晝之夜，畸零地工作室把名為「古蹟燒」的攤車推到街上，以傳統小吃雞蛋燒結合被燒毀的老建築視覺意象，並透過文資知識的故事性傳播與街頭表演，試圖喚起大眾對文化資產的保存意識。攤車前後造訪這座島上多個遭焚毀的古蹟遺址，也積累了不少參與民眾與地理上的製圖（mapping）里程。古蹟燒攤車造訪苑裡數次，也因緣際會之下結識了在地組織「掀海風」，共同為市場的毀敗與再生付諸一次實踐行動。

> 2018 年 9 月 14 日，一場無名大火將苗栗苑裡的百年公有零售消費市場燒成斷壁殘垣。2019 年 10 月 18 日，應運而生的演出《苑裡好人》在市場遺址首演。從苑裡市場的燒毀到演出，前後相隔一年許。然而，在這中間被省略的，卻是許多在看最後演出的觀眾所難以觸及的：受戶的何去何從、歷史建築身分的登錄、多場意見分歧的民眾參與式討論工作坊與公聽會難以定案的重建計畫、演出團隊的籌備規劃等，都在這僅僅一年的時間中接踵而生。

> 為苑裡市場「製造事件」的同時也是向社會大眾的一次嗆聲。苑裡市場是重要的，所有自燃的古蹟是只要說消失了就沒了。《苑裡好人》作為對已逝古蹟的一次存檔，才有可能讓後繼者體會到，歷及其物質性遺留是需要守護的。而這個守護，更是需要一再被實踐。（文｜謝鎮逸）

• 觀察報告

謝鎮逸〈不好不壞不在場的民主：畸零地工作室的文資關懷策略〉，全文請見
https://mag.clab.org.tw/clabo-article/ground-zero-and-cultural-heritage/

污痕結構學：城市記憶空間的建築文化學實驗

Reconstruction of Stain: A Transdisciplinary Experiment on the Space of Memories in Taipei and Berlin

我們該如何面對歷史遺留下來的「污痕」？又能賦予「污痕」怎樣可見的形構？本計畫將結合文化學、建築學以及身體理論的觀點重新回顧「髒污」的多重意義，並透過柏林的案例分析來探討一座城市能以何種方式來與「髒污」共處，而又能以何種形式將「歷史的污痕」標記在地景之上。是巍然聳立的高塔？或是低伏而開闊的平面？是堅實、沉重的石材？或是通透、一目瞭然的玻璃？本計畫將重新爬梳環繞在各種形式、媒材之上的意圖、想像與意識形態，並設計一套能讓公眾實際參與操作的「記憶構件」，透過不斷拆解、組合的過程，重新去思索如何展示「污痕」；如何與「歷史」共處，並發掘「記憶空間」的不同可能形式。

How should we face the "stains" of history? What sorts of form and structure can we provide for stains? This project combines concepts from cultural studies, architecture, and body theory to retrace the multifaceted meanings of "stain." By inspecting the case of Berlin, this project explores ways a city can coexist with "stains" and methods of tagging "historical stains" on the landscape. How can this be done? By erecting tall towers and monuments? Or open and rolling plains? By using sturdy, heavy mineral materials? Or transparent glass? This project investigates intentions, imaginations, and ideologies that surround different forms and mediums and designs a set of memory components that allow public participation and operation. By repeated deconstruction and reconstruction, this project reinspects ways of displaying "stains," coexisting with "history," and discovering different possibilities of "the space of memory."

REATORS

李立鈞 LEE Li-Chun

柏林洪堡大學文化學系博士。圖像研究者。於國立臺灣師範大學美術學系畢業後，負笈德國，於柏林洪堡大學修習藝術、圖像史（Kunst-und Bildgeschichte）與文化學（Kulturwissenschaft）。

LEE Li-Chun holds a PhD from Humboldt University of Berlin in cultural studies. His profession is iconology. After graduating from Department of Fine Arts at National Taiwan Normal University, he left for Germany to study art, visual history, and cultural studies.

吳耀庭 WU Yao-Ting

「Ursprung」建築圖書室負責人、「現場的建築史」資料庫創辦人。畢業於瑞士聯邦理工學院都市設計碩士。曾工作於宜蘭田中央聯合建築師事務所、嶼山工房，重視手做模型與現場考察。

WU Yao-Ting in charge of Ursprung Bibliothek and the founder of online dadabase Vers Une Architecture. He received his MA from Swiss Federal Institute of Technology Zurich in urban planning. He used to work for Fieldoffice Architects in Yilan and Atelier Or. He pays special attention to hand-made models and on-site investigation.

謝杰廷 HSIEH Chieh-Ting

舞蹈音樂研究者、音樂家、藝術家。 曾與音樂家大竹研、早川徹合作，其創作曾於臺北市立美術館、柏林 Galerie im Turm 展出。近年於德國從事音樂舞蹈研究，研究興趣涉及從現象學與文化技術觀點探察音樂與舞蹈的身體感、力動、記譜等

HSIEH Chieh-Ting is a dance music researcher, musician, and artist. He collaborated with Ken Ohtake and Toru Hayakawa. His works were exhibited in Taipei Fine Arts Museum and Galerie im Turm in Berlin. In recent years he engages in dance music studies in Germany. His research interests are in bodily perception, energy, and notation of dance from the perspective of phenomenology and cultural technique.

李立鈞、吳耀庭、謝杰廷，「污痕結構學：城市記憶空間的建築文化學實驗」，2019_

LEE Li-Chun, WU Yao-Ting, HSIEH Chieh-Ting, *Reconstruction of Stain: A Transdisciplinary Experiment on the Space of Memories*, 2019_

> 「污痕」既是概念的破題也是引線針，試圖鬆動「污痕便需要被去除」的認知，並指出污痕所製造出的另一個現實亦有它能夠開啟的討論空間。「污痕」的概念變成掌權者的政治話語權，並作為宣傳或迫害的工具，如納粹對猶太人進行的種族肅清；然而在戰後，納粹之於德國也同樣被視為污痕。以「污痕結構學」的觀點來看，李立鈞認為：「是要將『污痕』概念所開啟、關於城市／歷史記憶的問題意識複雜化，並非要做非黑即白的斷言，而是要把這之間的光譜、當中的複雜性和曖昧性，藉由各個學科進行探問。」在計畫期間的開放工作室活動日，三人取徑自建築、歷史、藝術創作領域，作為探討「痕跡」的多元案例，如謝杰廷所言：「關乎兩個物件／物質碰觸在一起的狀態，並在雙方的『身體』產生可見或不可見的痕跡。」

> 三人於柏林的田野調查著重觀察德國人如何去處理被標記為「污痕」的歷史痕跡，以及這些痕跡在當地被呈現時所訴諸的語彙，以建築方面來說是關乎形式與媒材的選定，其他可能性諸如檔案資料的整理、機構的成立或是藝術創作的展現。吳耀庭提到「痕跡」不只代表歷史的證據或差異，亦要進一步思考它被看見的驅力——是警示意味或是喚起同理心、是期待人們以中性的立場理解還是帶著某種情緒去看待。

> 「污痕結構學」視城市、紀念物與建築為可閱讀的記憶媒介，三人嘗試藉由舉辦「記憶組構工作坊」將抽象的記憶轉化為具體可組構的積木元件，其造型與樣式靈感多取材於計畫調研中歷史上曾實際運用於建築的材料，如玻璃、金屬、木材、石製等，再結合其質地差異所賦予的光滑、粗糙、尖銳、負重等不同感受，象徵討論歷史事件時會涉及的敘述形容，有較具體地以多出現於西方建築當中的記功柱為原型的垂直造型對應「榮耀事蹟」；也可能是指向某種身分、行為或是某種狀態，像是取椎體或塊體的幾何造型比擬為「受壓迫的少數」、「異議者」、「歧視」或「被修改的」；另外，也會以兩種材質去對應同一詞彙，以「戰爭」為例，分別有具份量感與壓迫感的實心混凝土圓球，以及可以對

半打開、作為容器置入其他元件的空心透明圓球等，藉由材質語彙的安排加強描述性。謝杰廷說道：「當這些積木代表記憶的時候，這層關係便不會只關乎空間跟建築語彙。」（文｜黃鈴珺）

● 觀察報告

黃鈴珺〈一面以城市 / 歷史記憶為材料的百納旗，談李立鈞、吳耀庭、謝杰廷的「污痕結構學」〉，全文請見
https://mag.clab.org.tw/clabo-article/reconstruction-of-stain/

三人更多在柏林田野調查的見解，請見〈關於柏林紀念物的 17 則筆記〉
https://mag.clab.org.tw/clabo-article/17-notes-on-berlin-memorial-objects/

頹傾城市
The Ruined City

此影像計畫製作了 部關於城市的錄像裝置作品，表現恍惚的空間、景象與記憶。城市作為一個群體所建構之景觀，其時間軸參雜著逝去的故事，正在進行的時間以及尚未到來的願景。

影片內容將暫時規劃幾個子題，並視收集資料過程調整合併或增減。本計畫將於創作完成後在空總創新基地中尋找合適之空間發表展出。

This project intends to make a video installation about a city, displaying blurry spaces, scenes, and memories. As a scenery constructed by a group of people, the city is intertwined with fleeting stories as well as ongoing moments and visions that have not yet arrived.

The content of the video will be temporarily separated into a few sub-themes that will be adjusted according to data collection and it will eventually be presented in a suitable space within Taiwan Contemporary Culture Lab.

CREATORS

張立人 CHANG Li-Ren

1983 年生於臺灣臺中，畢業於國立臺南藝術大學造形研究所，創作多以錄像裝置、觀念計劃和動畫為主，擅長用敘事的手法建構出介於想像與現實之間的虛擬世界。

Born in Taichung in 1983, CHANG Li-Ren graduated from Graduate Institute of Plastic Arts at Tainan National University of the Arts. His works mainly consists of video installations, conceptual art and animations created from his story-telling techniques, featuring a virtual world that exists somewhere between imagination and reality.

「向大陸沿海的同胞們廣播」
“and I’d to express my feelings to my compatriots on the mainland coast.”

▌張立人，「頹傾城市」，2019_

CHANG Li-Ren, *The Ruined City*, 2019_

> 《FM100.8》為張立人邀請成媛與芮蘭馨共同參與的計畫，以關注空軍總司令部舊址（現為臺灣當代文化實驗場 C-LAB）和旁鄰的正義國宅為起始，不僅在於這些場域獨特的氛圍，也包含過去至今存在於其中的個體，以及持續變化的空間狀態。張立人和成媛初次談及構想時，曾提到當前的社會狀態猶如「不同時空的結合體」，人們所感知的現實來自於被提供的環境，諸如隨著時代更迭所遺留、堆疊的具體痕跡或是抽象的時代意識，人們便據此構建出自己的生活樣態。然而「那些經歷時代意識洪流沖刷過而仍活著的人」亦因著這樣的生存狀態迫使出某種驅力，建構出某種想望，卻無從知曉該望向什麼樣的彼方。

> 計畫以《FM100.8》為命名，象徵著歷史流逝中一段段時間切片，如同廣播調頻的經驗，並以此為靈感串起創作的敘事架構。廣播運作是透過電信訊號接收與播送訊息，收音機則是各個經驗接收與播送的媒介。若說無線電波與腦波有某種關乎頻率傳遞反應的相似之處，廣播波段或能視為與意念、能量相通的存在，甚至得以指向任何東西恰如張立人所言：「其實就有點像是心電感應，我喜歡用無線電來形容，就是可以感到對方在想什麼。」他對於短波廣播尤感興趣，當它成為專制政權下，國家加強控制工具時，其特色是會覆蓋某些電台，用以避免人民聽到外國頻道，然後不斷地放送相的雜訊或者歌曲，且內容並未有任何表明身分、消息或任何可作為資訊識別的片刻，間感亦在如此的重複之中喪失。

> 在敘事編排上，張立人以「巨人」、「海市蜃樓 / 桃花源」、「迷宮」與「多重視覺」四個子題交織概念層層描繪正義國宅和空總的島嶼意象、金門與廈門的島嶼經驗與舊時的戰地氛圍，反映出冷戰時期黨國體制下的時代意識，以及個體與集體之間的種種存在狀態，並交錯著來自成媛和芮蘭馨所書寫的與之呼應的生命經驗或特別觸動的記憶。由三人共同完成的文本，有著闡述的、宣揚的、獨白的、

顧的敘述口吻，卻不明確點名是屬於誰的故事，並剪輯著在這段期間，拍攝自空總、正義國宅、金門等地的景象、勾起回憶的場所與詩意的象徵畫面。（文｜黃鈴珺）

觀察報告

黃鈴珺〈光波粼粼中映現的意識餘像：張立人、成媛、芮蘭馨《FM100.8》〉，全文請見
https://mag.clab.org.tw/clabo-article/the-ruined-city_fm100-8/

投聲計畫
Transonic 2020

隨著當代科技發展的進步與普及，在每個畫面與 pixel 都斤斤計較的時代，人在視覺上致力地用飽和的色彩、幀數更新率來追求擬真、還原度，甚至利用視覺神經元感受的極限，來創造 3D 的環境、景深的重新演算。但在聲音、聽覺上，人渴望的是什麼？對我來說，波場合成 (Wave Field Synthesis，簡稱 WFS) 的聲音合成技術，是一道讓聲音與空間關係，進入到另一個層次的大門。

這個計畫經由製作、研究、創作三個階段達成。試圖延伸過去聲音裝置作品《脆弱的透明》、《何處》所處理的聲音與空間的狀態、視覺與聽覺的討論，並加入 2018 年駐村發展，到 2019 年白晝之夜發表的百人行動裝置演出《聲音路徑——日行東西》的概念，來重新思考當代聲響再現的演進上，所形塑出的聆聽模式，另外我也好奇當聽覺接近真實、超越現有媒體認知的展現方式時，人對於聲音親密感所產生的反應。

Contemporary technology improves and popularizes every day in the era of particular demand for pixel resolution. People energetically pursue verisimilitude and reduction degree through saturated colors and higher refresh rate of frames. They even create 3D environments and perception of depth by carrying their visual neurons to extremes. In terms of sound and auditory sense, however, what do we actually long for? As far as I'm concerned, the technology of Wave Field Synthesis (WFS) elevated the relationship between sound and space to a higher level.

This project proceeds in three phases, including production, research and creation. I try to continue the discussions about sound and space as well as vision and auditory sense in my sound installations *The Resonance of Fragile* and *Where*, and apply the concept picked up from *Sound-path*, a production developed in 2018 when I was an artist-in-residence and it was presented at the Nuit Blanche Taipei 2019, so as to rethink the mode of listening shaped over the course of contemporary sound representation. Besides, I'm very curious about how people would react to the sound intimacy they experience when it appeals to their auditory sense in a way that transcends all existing interface.

REATORS

吳秉聖 WU Ping-Sheng

以聲音為創作思考中心的藝術工作者。2015、2016 年入選「數位藝術獎」、參與《Render Ghost》團隊拿下「數位藝術表演獎首獎」。2018 年獲選韓國駐村計畫。2019 年獲得文化部合作甄選至紐約藝術機構「三角藝術協會」駐村。2019 年六月代表 C-LAB 赴龐畢度中心 IRCAM 實驗室參訪交流，於臺北白晝之夜發表聲音演出作品。2020 年受邀臺中歌劇製作《無光風景》。並於新媒體藝術企劃團隊「噪咖事務所」擔任音樂統籌。

As a sound artist, WU Ping-Sheng was nominated for the "Digital Art Awards Taipei" in 2015 and 2016. He was also a member of the production team of *Render Ghost*, the first prize-winner of the 6th Digital Art Performance Awards. He worked as an artist-in-residence in Korea (2018) and New York (2019 at the Triangle Arts Association), visited the Centre Pompidouon's IRCAM on behalf of the C-LAB in June 2019, and presented his sound performance at the Nuit Blanche Taipei. In 2020, he took part in the production of *The Scenery of Little Light* at the invitation of the National Taichung Theater, and serves as the music and sound director for noiseKitchen, a team focusing on new media art.

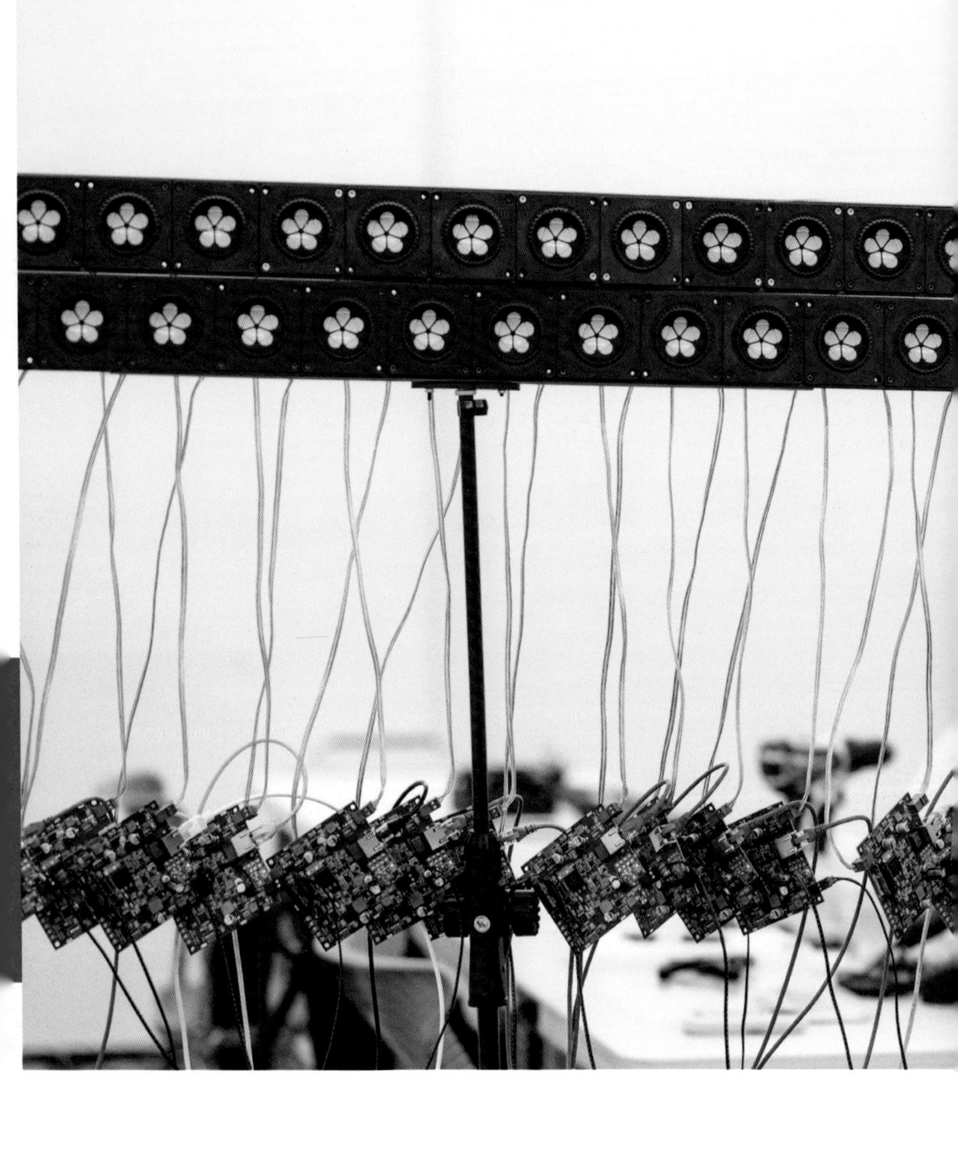

吳秉聖，「投聲計畫 2020」，2020_
WU Ping-Sheng, *Transonic 2020*, 2020_

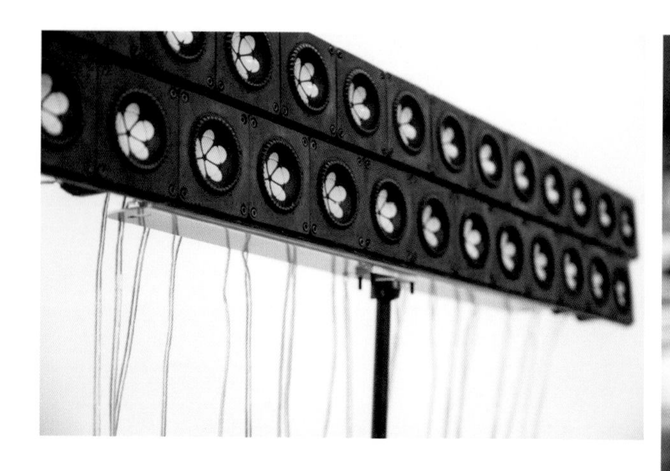

> 「投聲計畫」延續自吳秉聖過去創作所關注的面向之一：「移動的聲音路徑／聲音的位移」，整個計畫在六個月的期程內進行不同階段的組裝、測試與系統修正，因國內少有此種聲音系統的應用與研究，吳秉聖憑藉之前赴美國紐約州參訪 Curtis R. Priem 實驗媒體與表演藝術中心（Curtis R. Priem Experimental Media and Performing Arts Center）與法國巴黎 IRCAM 的交流經驗，從零開始打造 WFS 聲音系統，歷經「研發製作」、「進駐研究」、「創作發展」三個階段。但受到疫情造成國際原物料運送延宕，或是在製作與研究階段遇上諸多的狀態排除、測試，需依賴不同技術背景的專業人員協助，在進駐的尾聲能將系統應用於創作發展的時間相當有限。

> WFS 聲音系統的播放效果在吳秉聖的測試下，有兩種模式：投放聲音於空間中，聽者透過來回移動感受聲音發聲點的確切定位；又或是聽者可以固定站立於一點，透過控制介面拖移發聲點以感受聲點的位移。此外，聲點的投放位置，不僅可以於喇叭正前方，亦可以是在正後方當創作者投放不只一個聲點音源於空間中，聽者在有效聆聽範圍中，可以辨識出聲音的相對遠近位置。

> 在聆聽體驗上，以人聲效果最為容易感受到聲音投放的精確位置，再來是相對具體的聲音，例如特定樂器的演奏聲，這種奠基於原有聆聽經驗的聲音，最容易幫助聽者快速辨別出具有意義的聲音，像是抓到暗室中的微光般，透過該聲音微弱的差異變化，逐步感受到聲音在空間中的定點位置。（文｜馮馨）

● 觀察報告

馮馨〈聲音場景的取樣、錄製與再建置〉，全文請見
https://mag.clab.org.tw/clabo-article/creators2020-3/

自然復刻：臺灣 AI 數位風景
Duplication of Nature: Digital Scenery of AI in Taiwan

「自然複刻：臺灣 AI 數位風景」這個計畫旨在探究藝術家在創作中，以貼合現代科技文明發展的 AI 技術（人工智慧）作為擴充工具的使用，並如何在這個計畫中發展並構築「人機共創」的數位工作方法。以數位和自然的混種作為創作思惟架構的主軸，並將資料庫的構成以社群共用作為擴充的方法，取代傳統單一來源的資料庫內容，利用社群概念搜集並擴充臺灣各個地區已存在的真實自然音景並交付人工智慧作為數據庫整理的輔助者。發展進一步的可能切入至創作之中，探討藝術家在創作的中的地位和主體性，能否和具有學習能力的人工智慧共存的可能性。

This project aims to investigate how the artist employs the latest AI technology as their creative tools and how they devise the digital modus operandi of “human-machine co-creation.” Revolving around the hybridization of the digital and nature, this project constructs a community-based database as a substitute for any traditional database that has only a single source of contents. Specifically speaking, it utilizes the concept of community to collect and expand the existing natural soundscapes around Taiwan, and collates these data with the help of AI technology, thereby opening up new possibilities for artistic creation. In sum, this project will reveal artist's status and subjectivity in the creative process and their potential to co-exist with teachable AI.

REATORS

陳志建 CHEN Chih-Chien

畢業於臺北藝術大學科技藝術研究所，致力於藝術創作，以混種跨域為概念，取材於自然環境，討論奇觀社會中的人性思維，希望在極度科技及極度人文這兩個極端中取得和諧，利用「視覺、影音、裝置、文本、行為」共陳手法回應當代議題，2006 年獲得臺北美術獎。2010 年創立藝術團體「豪華朗機工」，持續跨國現地創作，駐村於洛杉磯、巴黎、釜山等城市。

CHEN Chih-Chien graduated from the Graduate School of Art and Technology, Taipei National University of the Arts. Drawing inspiration from the natural environment, he not only orientates his artistic practice towards hybridization and boundary-crossing, but also deals with the humanistic thinking in the society of the spectacle, insofar as to strike a harmonious balance between the two poles of technologies and humanities. He tends to respond to contemporary issues by juxtaposing "vision, video, installation, text and behavior." He won the Taipei Art Awards in 2006 and founded the artist group LuxuryLogico in 2010. He continues to carry out site-specific creative projects in different countries and works as an artist-in-residence in different cities such as Los Angeles, Paris and Busan.

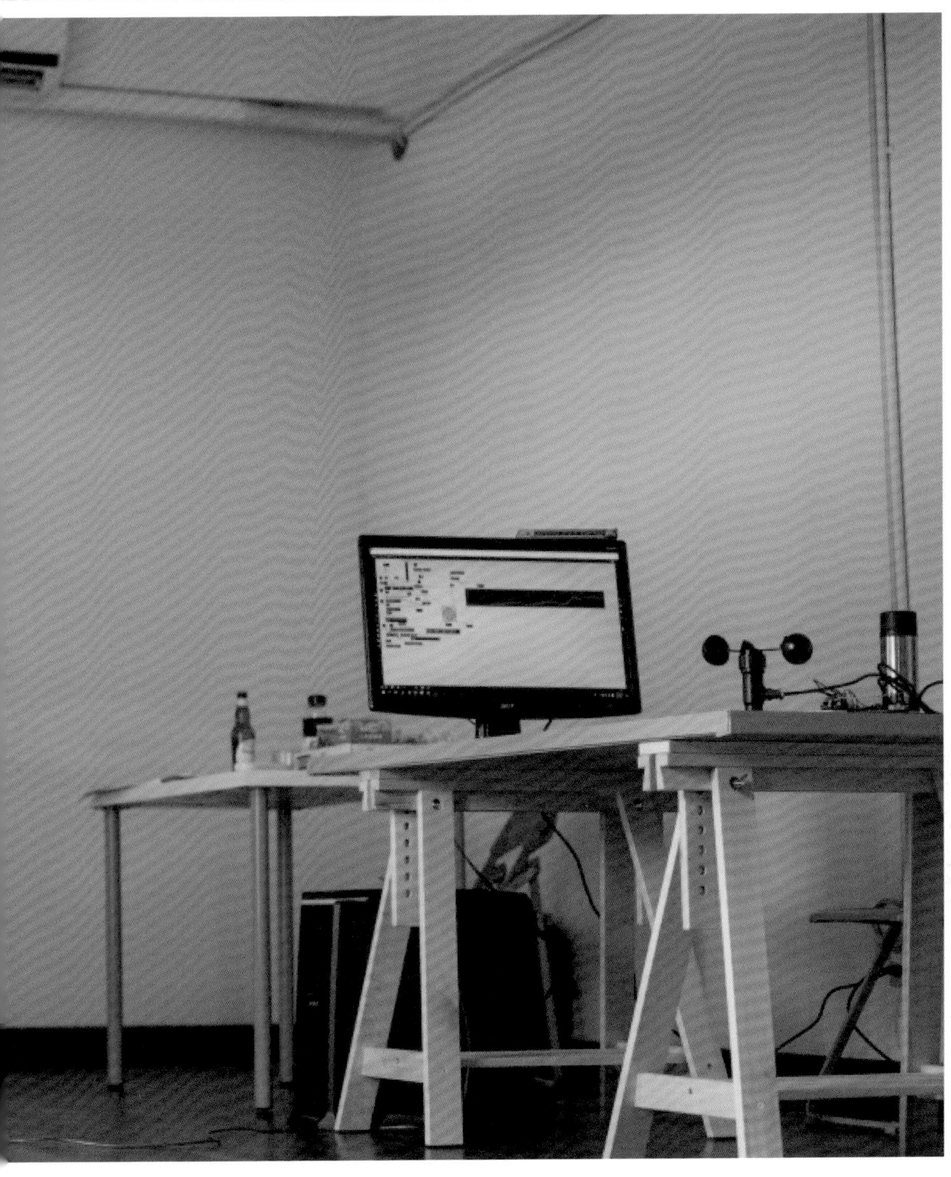

陳志建，「自然復刻：臺灣 AI 數位風景」，2020_

CHEN Chih-Chien, *Duplication of Nature: Digital Scenery of AI in Taiwan*, 2020_

> 讓我們設想這樣一個場景：在自然生態已被破壞殆盡的近未來，人們若要認識「自然」，便只能透過存取那些晚期人類世時期的田野工作者與 AI 所共構出的資料庫，來認識鳥獸草木之屬。在全球自然災害頻傳之際，這個末日感十足的景像似乎不再是難以想像的遙遠未來，而更像是有可能發生在有生之年得見的近未來。

> 讓我們設想另外一個場景：我們走入一個房間內，一面由光線與聲響組成的波浪湧來。仔細一看，是上百隻光條搖曳著，彷彿是一面芒花海般，間或伴隨著風刷過葉面的窸窣聲響以及斷斷續續的蟲鳴聲。這幅具有未來感的景象，可以說是前一個場景的具體化呈現：在自然持續被人類破壞的未來，甚至連「風」都將不復存在。此時，於末世倖存下來的人類若要理解「風」，這個他們未曾經驗過的自然氣象，將只能透過 AI 演算出依照過往紀錄所推測出的可能狀態，並以某種形式表現出來。因此，基於這樣的架構，我們所看到的光條運動軌跡，將會是 AI 基於過往某一地點「風」的累積資料所推算出的可面貌。而為了呈現出這個場景，首先要處理的是驅動燈條運動與音響震動的訊號如透過 AI 生成？接著，是如何調校機械動力裝置，使之呈現出藝術家心目中的「風應有的姿態？

> 在陳志建面向未來的計畫中，能感受到一個古老的意象，也就是諾亞那幾乎包含所有物種的基因舟。在諾亞的故事中，我們看到了一個經由毀壞而重建的世界，但這世界並不新穎，而是在耶和各從其類的律令中將世界還原（undo）為其應有的面目。因此，諾亞的時間是一種神話的、循環時間：一個被允諾的、可預期的「未來」。但陳志建的計畫與之不同，從開始的計畫到轉型的過程中不可預期總是現身，並要求著遭逢的人做出決斷：這意味著不再有律令、也不再有允諾，而是透選擇構建真正的、未知的未來。AI 工具也不只是將現世存有轉化為數據的封包，以便於在未來重

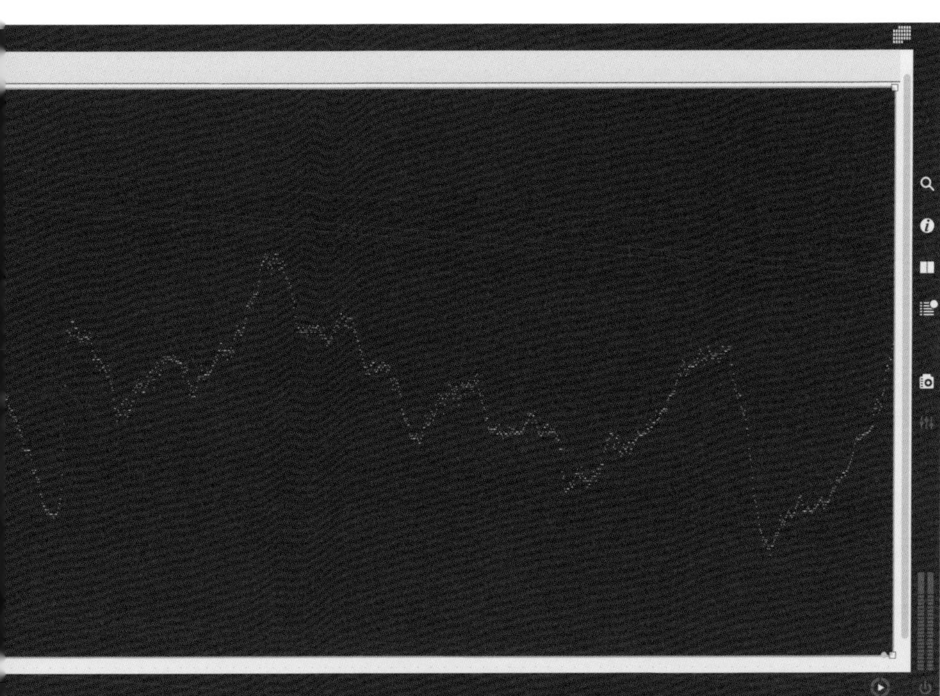

```
e-reuse):
ted(x,512,tf.nn.leaky_relu)
ted(outputs,256,tf.nn.leaky_relu)
ted(outputs,128,tf.nn.leaky_relu)
ted(outputs,64,tf.nn.leaky_relu)
ted(outputs,1,tf.nn.sigmoid)

n(tf.GraphKeys.GLOBAL_VARIABLES, scope='predi
```

```
AUTOGRAPH_VERBOSITY=10`) and attach the full output. Cause: converting <bound method Dense.call of
<tensorflow.python.layers.core.Dense object at 0x7f8ba9694ef0>>: AttributeError: module 'gast' has
no attribute 'Index'
WARNING:tensorflow:Entity <bound method Dense.call of <tensorflow.python.layers.core.Dense object
at 0x7f8ba9694f60>> could not be transformed and will be executed as-is. Please report this to the
AutgoGraph team. When filing the bug, set the verbosity to 10 (on Linux, `export
AUTOGRAPH_VERBOSITY=10`) and attach the full output. Cause: converting <bound method Dense.call of
<tensorflow.python.layers.core.Dense object at 0x7f8ba9694f60>>: AttributeError: module 'gast' has
no attribute 'Index'
WARNING: Entity <bound method Dense.call of <tensorflow.python.layers.core.Dense object at
0x7f8ba9694f60>> could not be transformed and will be executed as-is. Please report this to the
AutgoGraph team. When filing the bug, set the verbosity to 10 (on Linux, `export
AUTOGRAPH_VERBOSITY=10`) and attach the full output. Cause: converting <bound method Dense.call of
<tensorflow.python.layers.core.Dense object at 0x7f8ba9694f60>>: AttributeError: module 'gast' has
no attribute 'Index'
WARNING:tensorflow:From /Users/ken/Dropbox (LuxuryLogico)/CREATORS創作_研發支持計畫/11_程式/
wind_value_generator_0909/wind_value_generator.py:43: The name tf.train.Saver is deprecated. Pleas
use tf.compat.v1.train.Saver instead.
```

一個當下；而是透過此一技術，在當下對未來的某種路徑設定與祈願，一種時間上的悖論：一個不會來臨的末世未來。（文｜沈克諭）

觀察報告

沈克諭〈不在此時：時差的潛勢〉，全文請見
https://mag.clab.org.tw/clabo-article/creators2020-1/

亞洲傳統音樂身體培訓交流計畫
The Bodies of Asian Theatre: Collaboration between Taiwan, Vietnam, Quanzhou (China)

江之翠劇場創團以來，在推動演員培訓與跨國交流上不遺餘力。從白桃房的舞踏工作坊、印尼宮廷／當代舞蹈工作坊，以及歐丁劇場工作坊等等，未來，我們希望能更聚焦於建立梨園戲的當代美學，因此提出亞洲音樂身體培訓交流計畫：首先，我們希望挖掘越南傳統藝術新的可能性，邀請越南籌歌（Ca trù）藝術家來臺帶領工作坊及相關講座。傳統藝術籌歌是越北特殊的一種歌詠詩歌形式，與臺灣傳統音樂南管，於 2009 年同一年被聯合國教科文組織列入為非物質文化遺產名錄，雙方皆從音樂表現形式，體現出各自文化所蘊含的哲學體系、信仰實踐及文學內容等，具有重要意義。

The Gang-a Tsui Theater has spared no effort to promote actor training and international exchange since its establishment, such as the Hakutobo Butoh Dance Workshop in Japan, the Indonesian Court Dance/Contemporary Dance Workshop, and the Odin Teatret Workshop. We expect to give a sharper focus on the contemporary aesthetics of Liyuan opera, and ergo we propose this exchange project. First of all, we plan to explore new possibilities of Vietnamese traditional art by inviting Ca trù artists to lead workshops and associated lectures in Taiwan. Ca trù is a special form of chant poetry exclusive to Northern Vietnam, which has been included in the UNESCO Intangible Cultural Heritage Lists in 2009 together with Nanguan, a style of traditional Taiwanese music. The significant meanings for the philosophy, religion and literature of the two cultures respectively find vivid expression in these two forms of art.

REATORS

江之翠劇場 Gang-a Tsui Theater

江之翠劇場」創立於 1993 年。近年重
製作包括 2020 年於臺灣戲曲藝術節重
《朱文走鬼》、2018 年復團之作《行
洛津》演出後隔年獲選參與法國外亞維
藝術節。創團團長周逸昌先生 2016 年
印尼求藝途中驟然離世，現由資深團員
佳雯及魏美慧在 2017 年於臺北市重新
案，並擔任正副團長一職。力求浸潤於
管之美的同時，延續前團長南管現代化
理念，致力賦予傳統藝術當代精神。

Founded in 1993, the Gang-a Tsui Theater's recent primary productions included the remaking of *Zubun Elopes with the Ghost* at the 2020 Taiwan Traditional Theatre Festival and *Passage to Lo-Jin*, a theatrical piece delivered in 2018 after its regrouping, which was also invited to the Festival OFF d'Avignon in 2019. The founder of the theater ZHOU Yi-Chang died on his way to Indonesia for arts in 2016. Its senior members CHEN Jia-Wen and WEI Mei-Hui re-registered the Gang-a Tsui Theater in Taipei in 2017, and now serve respectively as the director and deputy director. They not only strive to promote the beauty of Nanguan, but also perpetuate the former director's ideal of modernizing Nanguan and imbuing traditional art with contemporary zeitgeist.

▌江之翠劇場，「亞洲傳統音樂身體培訓交流計畫」，2020_

Gang-a Tsui Theater, *The Bodies of Asian Theatre: Collaboration between Taiwan, Vietnam, Quanzhou (China)*, 2020_

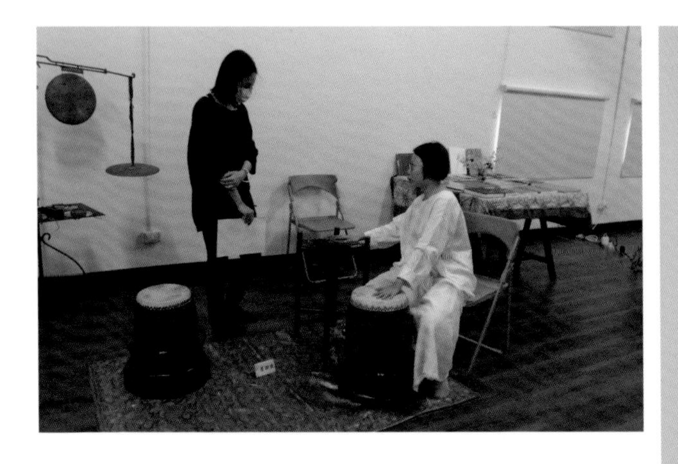

> 江之翠劇場的進駐計畫「2020 亞洲傳統藝術音樂身體培訓交流計畫」，除了延續團內南管與梨園戲的長年訓練以外，另一初衷旨在 2020 年下旬邀請越南傳統曲藝「籌歌」樂師訪臺，進行培訓工作坊與講演、交流。回頭看看劇團所提交的計畫其名—— 2020、亞洲、傳統、藝術、音樂、身體、培訓、交流、計畫——幾近完整構成了江之翠劇場進駐理想的九組關鍵詞。

> 現代化要實現，交流、混種即是最大亦為最直接的取徑。無論是身／聲訓練，抑或是與他團、邀請導演共製演出，皆是透過各種與外部的借力使力而形成。當然，向外借力的先決條件無非是先要有紮實穩定的基礎。在此，江之翠劇場所形構出的「亞洲」表演面貌即是在此般的交互作用下走出別於一般傳統表演的型態。

> 源自北越的傳統宮廷表演「籌歌」，確切出現年代已不可考，但 15 世紀已傳佈北越各地。文本形式以傳統越南詩體入歌，並含 56 種曲式旋律。演出陣容基本上由一名女性歌手與兩位男性樂師組成。歌手一邊敲擊拍板一邊演唱，咬字清晰並以特殊呼吸技巧、顫音與裝飾音等方式演唱。兩名樂師則分別彈奏三弦琴和讚美鼓。籌歌有時也會加入舞蹈，演出場合除了宮廷表演，也會用作祭儀典禮、競賽演奏或民間娛樂。20 世紀初期的多次戰爭，使越南飽受各方摧殘。在歷經大規模饑荒、1945 年越南八月革命、法越戰爭以降，籌歌曾一度瀕臨絕跡。當時甚至因政治壓迫，籌歌表演者以曾擔任官場藝妓而長久以來被扣上不貞潔之名。直到 2009 年，才被地方極力引薦予聯合國教科文組織，方才列為非物質文化遺產名錄，終為眾人所重視。

> 籌歌的琴、唱搭配如影隨形——一如南管。且其真聲技巧與頭腔共鳴的運作原理並非當地演唱傳統，卻與南管的真聲本嗓有著「異曲同工」之妙。雙方如此接近的經驗，也讓劇團思索如何向越南文化學習，將籌歌的發聲技術透過培訓工作坊納入訓練之中。（文｜謝鎮逸）

• 觀察報告

謝鎮逸〈2020 CREATORS「江之翠劇場」與「她的實驗室空間集」的觀察報告〉，全文請見
https://mag.clab.org.tw/clabo-article/creators2020-6/

芥面：正體中文版1.0
Intergrass: Traditional Chinese Edition 1.0

本計畫關注目前網路言論自由的挾制，並以臺灣應對中國官方在漢語文資訊傳遞與儲存的監督控管為例，參照過去臺灣與中國資訊交流的歷史，和香港反送中運動與武漢疫情的當下，網路資訊傳輸的現實需求與潛在困境，結合資訊科學與基因工程，提出以基因改造阿拉伯芥此一模式植物作為兩岸三地資訊交流以及語彙保存介面（芥面）的完整方法與教學示範，並作為未來國際間任一網路極權控管區域資訊無法自由傳輸甚或「斷網」情境下的匿名訊息傳遞、擴散、複製、儲存的格式，也企圖藉由本計畫潛在的後續行動「擴散全球生態影響不明」的基改生物之生態環境爭議，進一步帶動討論極端情境下資訊戰略的適切策略、倫理規範與自我審查。

This project uses the example of Taiwan's response to China's surveillance and control over Mandarin information transfer and storage to highlight the restriction upon the freedom of speech online. Considering several serious issues such as the actual needs and potential difficulties of online information sharing in the past history of information exchange between Taiwan and China, the Anti-Extradition Law Amendment Bill Movement in Hong Kong, and the COVID-19 pandemic, this project proposes genetically modified Arabidopsis thaliana as the interface for cross-strait information exchange and vocabulary preservation by integrating information science with genetic engineering and establishing a complete set of methods and teaching instructions. The interface will also serve as a format of anonymous information transfer, proliferation, reproduction and storage when online information cannot be transferred freely under the control of any potential online totalitarian regime or in face of Internet disconnection in the future. Moreover, this project attempts to evoke further discussions about appropriate information warfare strategies in extreme scenarios, ethical code, and self-censorship from the "unclear implications of genetically modified organisms for global ecology."

CREATORS

遠房親戚實驗室
Lab of the Distant Relatives

遠房親戚實驗室之虛擬空間永在，肉身成員流動。此次「芥面：正體中文版 1.0」由曹存慧、吳柏旻、楊克鈞、葛昌惠、陳韋臻等人負責執行。以物種論，非日萬物皆有靈，而是萬物皆為遠房親戚；以法律論，民法第 983 條有令，旁系血親在六親等以內者不得結婚，以人類趨避行為慣性而言，七等親成為遠房親戚定義的起點。六度分隔理論見證了社群時代的收斂，七等親以外的廣大田野是遠房親戚的發散磁場。實驗室，先講究效果，再研究不傷身體，並持續規劃一間對照室，繁殖永恆的思想發（渙）散。

The virtual space of Lab of the Distant Relatives exists forever, but the physical bodies of its members are fluid. For Intergrass: Traditional Chinese Edition 1.0, Theresa TSAO Tsun-Hui, WU Po-Min, YANG Ko-Chun, GE Chang-Hue, and CHEN Wei-Chen will be responsible for its development. Although the Theory of Species does not mention the pneuma of living creatures, it describes all organisms as distant relatives. From a legal perspective, Article 983 of Taiwan's Civil Law Code states that one should not marry a collateral relative by blood within the sixth degree of kinship. According to customary human avoidance behavior, distant relatives are relatives beyond the seventh degree of relations. The theory of the Six Degrees of Separation shows the convergence of the social community era, while relatives beyond the seventh degree are on the enlarged magnetic field of divergence. Our lab works firstly on creating the optimal impacts, and then studies to avoid physical risks. We are always preparing a control lab in contrast to the experimental lab, in which diverging/distracting thoughts procreated immortally.

英文
习近平
米兔

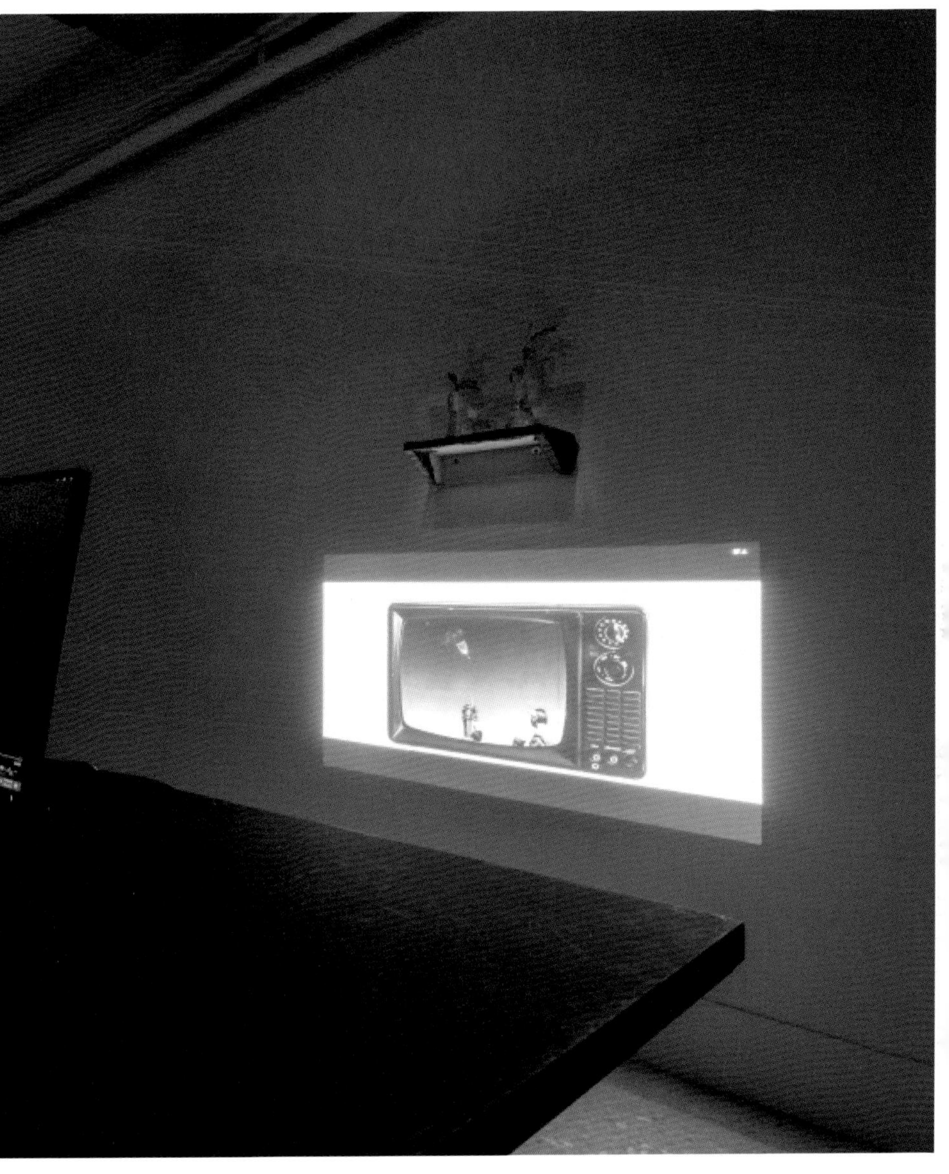

遠房親戚實驗室，「芥面：正體中文版 1.0」，2020_

Lab of the Distant Relatives, *Intergrass: Traditional Chinese Edition 1.0*, 2020_

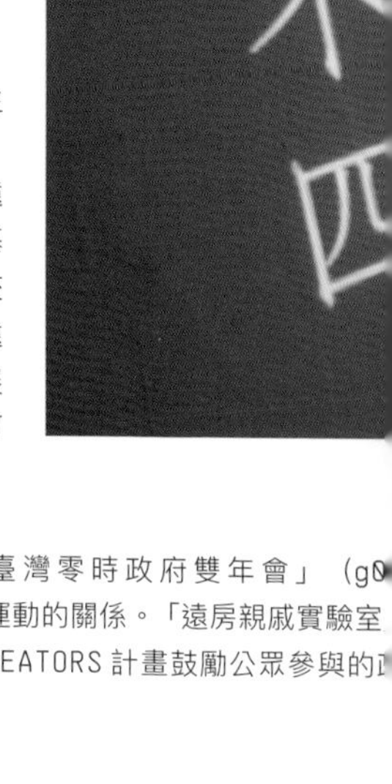

> 「遠房親戚實驗室」在作品《芥面：正體中文版 1.0》（簡稱《芥面》）中，團隊設想將基因工程技術加入未來資訊戰的反堵策略，設計將阿拉伯芥作為保留敏感詞彙的生物載體。然而，若把所有作品介紹遮住，這看起來就只是典型的植物分子生物學實驗，若以我們的話來說，就是「把文字資訊轉成核酸序列，訂購合成序列接進 pCAMBIA 接 YFP 當 marker，dipping 到阿拉伯芥後收 T2。」

> 「遠房親戚實驗室」內部存在跨領域機制，成員涵蓋生物學家、攝影師、程式設計師等。作品發想完成後，專案管理者會依工作內容將業務分給成員，接著便遠端運作，互不干涉。以作品《芥面》為例：發想及基因轉殖由曹存慧捉刀，待概念與作品有雛型後，才交給葛昌惠負責拍攝。進駐期間的開放工作室當天，「遠房親戚實驗室」成員依各自的專長與參與程度，發展出多種解說作品的方式，也聆聽其他領域的夥伴如何理解作品。

> 2020 年 12 月，他們把進行一半的《芥面》帶去臺南的「臺灣零時政府雙年會」（g0 summit），在那之前，還與世新大學的學生一起討論媒體與社會運動的關係。「遠房親戚實驗室的成員，原本便與開放原始碼社群有不少聯繫，這也讓他們與 CREATORS 計畫鼓勵公眾參與的政策不謀而合。（文 | 王順德）

• 觀察報告

王順德〈我的專業被你拿去跨界？來自生科人的文化實驗觀察報告〉，全文請見
https://mag.clab.org.tw/clabo-article/creators2020-7/

噥噥時間交換所
Murmur Time Project

走路草農藝團 2019 年舉辦「農閒藝術節」，思考日常生活中的季節更迭，與時間交換的狀態。延續「農閒藝術節」與「思箱計劃」的思路，配合 C-LAB 的地理位置關係，成立「噥噥時間交換所」，如果說「農」是種植技術、保存方法與土地記憶，那「噥」便是上述過程中的人與人的對話與交流。

農人在忙碌的農事之餘，是否有真正的農閒時間呢？還是一刻都閒不下來？農人的農閒都在忙什麼呢？城市裡的種植，算不算一種農夫？種植的渴望，期待的是什麼？「時間的交換—農閒體驗場」、「記憶的交換—空總農民曆」、「歷史的交換—蔬果發電所」三階段創作，將成為「噥噥時間交換所」的發表項目，在參與式的活動中與參與者交換不同日常的時間狀態。

The Walking Grass Agriculture organized the Farming & Arts Festival in 2019, inviting the participants to contemplate the alternation of seasons in their quotidian existence and the exchange of time. In coordination with the geographical location of the C-LAB and following the philosophies behind the Farming & Arts Festival and the project "Homesick Box," we established the Murmur Time Project. If "farming" refers to planting techniques, a preservation method and a memory of land, "murmur" is ergo the interpersonal dialogue and exchange taking place in the aforementioned process.

Do farmers really have spare time when they are busy with farming? Or, they cannot stop working? What do they do in their spare time? Can we count the people who grow plants in cities as farmers? What do people expect when they're keen to grow plants? The results of the Murmur Time Project will be presented in three phases, including "the exchange of time—farming experience site", "the exchange of memories—the Chinese lunar calendar at the C-LAB", and "the exchange of history—power plant of fruits and vegetables", allowing the participants to experience different temporal states in participatory events.

REATORS

走路草農/藝團
alking Grass Agriculture

農/藝　農一藝　農藝　農|藝　在農事與藝術之間　有各種可能性」，走路草農藝團成立於 2014 年，為一視覺藝術、設計和策展團隊。本計畫成員來自新媒體藝術、藝術史背景、生物科學等領域，強調駐地體驗與觀察，對於民藝的學習、遷徙與生成充滿興趣，結合地誌學、考現學與物質文化的研究，轉化農事經驗，形塑自己的藝術方法。近年來關注生態環境、都市變遷與性別議題。

There are infinite possibilities between farming and art. Founded in 2014, the Walking Grass Agriculture is a team engaging in visual arts, design and curating. The team members of this project come from the fields of new media art, art history and biological sciences. They set great store by residency experience and observation. Deeply interested in the learning, migration and generation of folk art, the members not only apply topology, modernology and material culture studies, but also transform their farming experience into their sui generis artistic practice. In recent years, they've focused on biological environment, urban change, and gender issues.

▌走路草農 /藝團，「嘌嘌時間交換所」，2020_

Walking Grass Agriculture, *Murmur Time Project*, 2020_

> 走路草的創作宗旨為「治本於農，以農入藝」，在這次 CREATORS 進駐計畫——「噥噥時間交換所」中，走路草在 C-LAB 打造了一個提供對話的場域，在都市分享農村的農忙、農閒的狀態，將農產地與農忙的紀錄帶入，講述農產品生產與醃漬製作等相關知識，將食用的體驗帶入計畫之中。參與者與藝術家從單一食物開始，延伸到食物生產地、製作方法等面向，藉由農產品交流彼此的生活與記憶，連結到產地之地緣脈絡，產生新的關係網絡。

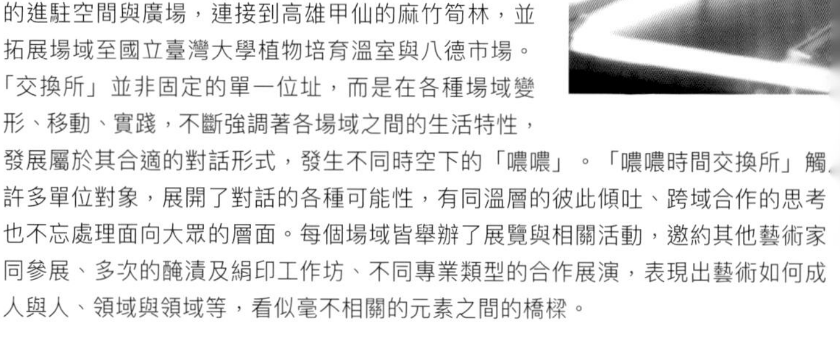

> 「噥噥時間交換所」的場域想像，是應用 C-LAB 內的進駐空間，藉由對話的形式，與藝術家、藝術行政、藝文單位、藝術媒體等不同對象，進行的資訊交流練習。短短五個多月來，「噥噥時間交換所」已從 C-LAB 的進駐空間與廣場，連接到高雄甲仙的麻竹筍林，並拓展場域至國立臺灣大學植物培育溫室與八德市場。「交換所」並非固定的單一位址，而是在各種場域變形、移動、實踐，不斷強調著各場域之間的生活特性，發展屬於其合適的對話形式，發生不同時空下的「噥噥」。「噥噥時間交換所」觸許多單位對象，展開了對話的各種可能性，有同溫層的彼此傾吐、跨域合作的思考也不忘處理面向大眾的層面。每個場域皆舉辦了展覽與相關活動，邀約其他藝術家同參展、多次的醃漬及絹印工作坊、不同專業類型的合作展演，表現出藝術如何成人與人、領域與領域等，看似毫不相關的元素之間的橋樑。

> 走路草常開玩笑說他們是 2020 年 CREATORS 計畫中最低科技的一組，然而如果用實驗研究的度來看，或許這些生命的故事才是最無法預期，且充滿可能性的，如同具備歷史縱深的釀酒與漬技術，必有其動人與超乎科技的存在必要。（文 | 王瑀）

觀察報告

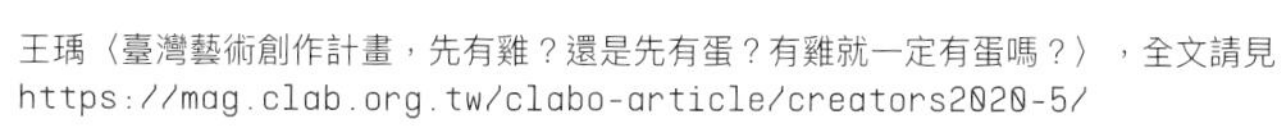

王瑀〈臺灣藝術創作計畫，先有雞？還是先有蛋？有雞就一定有蛋嗎？〉，全文請見
https://mag.clab.org.tw/clabo-article/creators2020-5/

如果，家族旅行最終章：研究與創作實驗發展計畫

If, a Final Chapter for a Family Trip. (Research and Experiment Project)

「如果，家族旅行最終章」由田野調查、書寫——聲響實驗、階段性發展所組成。透過一份 「民謠緊急調查報告書」的想像為起點，從僅有的文字紀錄中，如何能重現相互黏著的文辭與旋律以及隱匿其後的「幽魂」？這是一趟探問與試煉的旅程，旅程中將重新思考檔案中的既存卻又模糊之邊界。而若相簿是某種私歷史的檔案，那相簿中的回憶與誤憶，可以如何演繹？本計畫以臺灣為中心，延伸往南、北之想像，創作「想像旅行」的對話，藉由採集當地居民對於歌謠的記憶，將文字或旋律重新賦予想像與意涵，並將多首歌謠匯集、組裝成一具劇場性的「最終篇章」，透過不斷地重新組合、組織、組裝的歷史、地理時空交換誌，將某種介於已過去之未來－當代－尚未抵達的未來的複數時間軸上無止盡循環的記憶，讓「最終篇章」可在不同空間場域進行各種形式的「演繹」。

This project consists of field survey, writing-sound experiment, and phased development. Treating an imaginary "investigation report of endangered ballads" as the point of departure, this project attempts to represent the interdependent lyrics and melodies as well as the "phantoms" behind them through the exiting written records. This is a journey of exploration and experimentation, on which we will re-examine the existing yet blurred boundaries within the archives. If photo albums are archives of private history, how can we interpret the memories and false memories they hold? Centering on Taiwan and stretching our imagination south- and north-bound, this project creates a dialogue of an "imaginary trip" by collecting locals' memories of ballads and giving new aura and connotation to the lyrics or melodies of ballads. Then, it complies and assembles several ballads and transforms them into a theatrical "coda." By virtue of the exchange records of history and geography that are continuously recombined, reorganized and re-assembled, this project tries to transform the endless loop of memories along the plural timelines betwixt and between the past future, contemporary, and upcoming future, thereby allowing the "coda" to be interpreted in various forms and venues.

REATORS

她的實驗室空間集
er Lab Space

初起於對空間的多層想像，進而從其發
見、實驗、創作，並融合多元創作形式，
採集日常小事、生活環境、歷史記憶為養
分，開展並衍伸更為豐富的面向，探觸其
多變的位置。團隊以「空間」做為創作發
展的主要題材要素，從多方面來探索空間
在時空裡的存在樣態與形貌，期待更多元
的實驗與跨領域合作創作。劇團針對空間
與環境著手調查研究，試圖在戲劇的框架
中將文本或戲劇中的「角色」置入適合的
現實場域或環境，找到兩者之間可以對話
的彼此交流空間，亦企圖將角色與不同的
建築體產生某種歷史感與生命經驗的對
話，此「空間」因而從實體建物的操作與
想像，推展至精神層面等的開啟與考掘。
期望未來將核心計畫更以不同領域創作
者、多樣性媒材與多元實驗性空間作為持
續發展基礎，透過邀請不同類型領域的創
作者共同工作，引入不同創作思維，建構
屬於「她的實驗室空間集」的各種可能。

Initiated from multiple imaginations about space, Her Lab Space draws on the living environment, historical memories and trivia of everyday life to conceive, experiment, and create in diverse forms, thereby developing and exploring extra dimensions and probing into the fickle nature. Treating “space” as the primary subject, the team looks into the modalities and appearances of space from multiple perspectives, and expects to engage in more diverse experiments and transdisciplinary collaboration. Starting by surveying spaces and environments, the troupe tries to appropriately situate the “roles” from texts or theaters in real spaces, and find the possibilities for exchange and dialogue between the roles and the venues. It also seeks to create a sense of history and a dialogue of life experiences between roles and different constructions, so as to transmute the “space” from a practice and imagination of physical construction into a source of spiritual enlightenment and discovery. Based on transdisciplinary collaboration, media diversity and multiple experimental spaces, Her Lab Space plans to invite artists from different fields to work together, introducing innovative thinking and opening up possibilities.

她的實驗室空間集，「如果，家族旅行最終章：研究與創作實驗發展計畫」，2020_

Her Lab Space, *If, a Final Chapter for a Family Trip. (Research and Experiment Project)*, 2020_

> 「她的實驗室空間集」的計畫「如果，家族旅行最終章」，從陳侑汝與區秀詒兩位藝術家的個人記憶與家族經驗，結合祖輩的故事、各自造訪過的洞穴、地方歌謠的流變，形構出亞洲的 ____ 部署。陳侑汝於 2017 年前往日本山口縣秋吉臺駐村，隨團參訪了景觀特殊的「秋芳洞」，並無意中得知歷經日殖時期的祖父在戰後開放出國觀光時也曾造訪過。而在距離秋芳洞 4,537 公里外、馬來半島的「金龍洞」，洞窟中神秘的古老虎化石也跟地方傳說與國族精神象徵有著近親性。區秀詒的外祖輩，亦因被口軍羈押的記憶，而痛恨日木人與臺籍日兵……大歷史與個人史、地緣與地質學藉著時間地層的出土與考掘，由點、線、面逐漸串聯出可能的 ____ 敘事。

> 2020 年 9 月 19 日的 CREATORS 開放工作室日，團隊呈現了一場小型展演「山穴的野狼眠夢，山豬的島嶼流亡」，做為該計畫的階段性集結。一位名叫「楊菊」的祖輩，回憶起戰時的馬來亞，口述日軍如何羈押她的丈夫並強行帶走。多年後的另一頭，同樣也是騎著腳踏車的一位臺灣人，在駐村時想要出發去一個叫「美爾」的小鎮，一路上的騎車時間與距離空間詭異地起了皺褶，裂縫處跑出了一隻路過的熊。老派電視廣告的主題曲響起——原來賣的是臺灣經日本本田（Honda）授權，以 CB125 為原型推出的經典打檔車野狼 125。這些林林總總的敘事之「點」，將透過藝術家之手，重新以「線」的串聯、縫合，最終重組成多維的言說之「面」。而這般「點—線—面」的文本連連看，正是剪裁及縫紉技術之實踐標的。

>

在區秀詒近期的個人計畫或「她的實驗室空間集」團隊計畫，開始更多關注老虎形象及其名字在各種跨國歷史、文化與視覺史上的流變，各種以哈力貓（ㄏㄚ　ㄌㄧˋ ㄇㄠ、ハリマオ、ha-ri-mau）之名的產物仿若多如東亞系譜中的貓科家族。然而，本次計畫中所提及古老虎化石洞穴的所在之地——近打河流域（Kinta Valley）當地對老虎的稱謂，至少在一百年前還有另一種名字：「Berolak」

據說在穆罕默德時代以前，有個人抓到一隻幼虎並帶回豢養，馴良的 Berolak 跟主人同住，直到主人離世後才回到了叢林裡，並長大成為體型巨大的老虎。當人們聽到它如雷的吼叫聲，聲音可穿透並迴盪於城都以北的珠寶區（Chemor）和以南的華都牙也（Batu Gajah）。若在旱季時聽到吼聲，必定是下雨十五天的徵兆。對同一事件的指稱仍然還在不斷出土中；而藝術家的使命必然也是持續去發掘陰性敘事與知識。（文｜謝鎮逸）

● 觀察報告

謝鎮逸〈2020 CREATORS「江之翠劇場」與「她的實驗室空間集」的觀察報告〉，全文請見 https://mag.clab.org.tw/clabo-article/creators2020-6/

採集者與版畫家——的確是存在於二十世紀
Collector & Printmaker in 20th Century

本計畫將追尋兩個身分迥異的神秘人物。第一，來自「自由畫社」的版畫家。他造訪各地，通過故事、身體轉化、構圖，帶著群眾雕刻，將發生於各鄉鎮的「白色故事」帶回故里。他延續1940年代的洞見，認為版畫有可通過「複印」而大量擴散的優點。於是，帶著大量群眾創作的「自由畫室」，將於下半年移走於全臺各地，邀請當地民眾重探自身鄉鎮消逝的幽魂，同時與他地之人交換視野，使白色的反思交互共鳴。第二，一位四處採集的「博物學者」。他四處探訪、繪製地圖，廣泛採集，最後將一具具的「叛逆者」轉化為剝製標本或標本箱，安置在「國家標本室」。他通過民間在「祭改」時作為「身體」替代物的「草人」，通過拼貼、謄寫歷史檔案而轉換成一具具剝製標本或標本箱，集中展出。我們想重新記憶或想像這群「也有快樂的、苦痛的、努力的，最少是在世上有做事的人」，然後在二十一世紀的時刻，將這些痕跡，通過創作，重新安置回空蕩的鄉鎮記憶。

The project revolves around two mysterious figures with totally different identities. The first is a printmaker from the "Free Painting Society." He visited different places where he led locals to create prints via stories, physical transformation and compositions, and then brought the "White Terror stories" unfolding in these places back to his hometown. Following the insight of the 1940s, he recognizes that prints have the merit of proliferation by "copying." Therefore, being about to drift around Taiwan with his "Free Atelier" in the second half of the year, he will not only invite locals to revisit the waning spirit of their native places, but also exchange horizons with them so as to stimulate reflections on the White Terror. The second figure is a natural historian who visited many places, drew maps and assembled an extensive collection. Finally, he turned the "rebels" he collected into specimens, and placed them in the "national specimen gallery." Specifically speaking, he turned the "scapegoats" in the ceremony of "avoiding mishap and adversity" into specimens by means of collage and historical archiving, and then exhibited them as a whole. We seek to re-memorize or imagine this group of people who were "blithering, suffering, striving or at least contributing to this world", and, by dint of our artistic creations in the 21st century, try to relocate these traces back to the vague memories of our hometowns.

CREATORS

安魂工作隊
Libera Work-Gang

「安魂工作隊」成立於 2018 年，是由一群有共同志向的藝術家、研究者、教育工作者、群眾所組成。工作隊的核心精神，是關注「戒嚴時期」消逝的抗爭者。而他們的亡佚，不只使身影消失在當代記憶，也使當代人喪失了一代砥礪彼此走入公共的典範——換言之，我們喪失了「魂」。「安魂」的實踐也就包括兩個向度：第一，讓這些人的身影重新在當代浮現。第二，讓當代人找回自己的力量，不因為主觀的保守，誤判了客觀上擁有的更大力量。我們期望通過兩向度的工作，從下往上，橫向連結、將當代群體的「魂」給重構出來。

Founded in 2018, the Libera Work-Gang is comprised of likeminded artists, researchers, educators and the masses. Concerns over the dissidents in the martial law period serve as the core philosophy behind its operation. The demise of these dissidents implies not only the disappearance of their physical bodies in contemporary memories, but also the loss of a paragon capable of galvanizing us to care about social issues and public affairs. In other words, we lost the "soul." As a result, the practice of the LIBERA WORK-GANG is twofold: (1) to allow these figures to reappear in the contemporary world; and (2) to help contemporary people muster their energy and prevent them from underestimating their objectively greater strength due to their subjective conservativeness. Using such a two-pronged strategy, we attempt to reconstruct the "soul" of contemporary people in a bottom-up and horizontally connected manner.

安魂工作隊，「採集者與版畫家——的確是存在於二十世紀」，2019-2020_

Libera Work-Gang, *Collector & Printmaker in 20th Century*, 2019-2020_

> 安魂工作隊是在現今從事白色恐怖記憶相關工作的團體中相當特別的一支隊伍，以「去臺北中心化、進入縣市鄉鎮」為準則，開展工作。他們的足跡廣佈，在許多鄉鎮甚至馬祖、小琉球，他們都曾前往。但他們的工作方法也很彈性，以「故事－身體工作坊－草人／版畫／其他創作」三階段工作坊為基礎，因應著不同地區的歷史及空間，甚至參與者屬性，適時做出調整。

> 三階段的工作坊並不是憑空而生，由林傳凱打頭陣，介紹與聽眾能產生連結（地域、職業等都有可能）的政治受難者，接著由林子寧帶領身體工作坊，以循序漸進的方式引導民眾進入感知與創作的狀態，並拍攝照片作為創作素材，最後才進入版畫轉印、草人或其他形式的創作。這樣設計最大的作用是民眾不必有任何基礎，幾乎完全零門檻。你不需要擁有很多白色恐怖的先備知識，不需要有表演藝術基礎，也不需要有繪畫基礎或是高深的構圖知識，你就可以參與安魂工作隊的工作坊，並在最後完成一個創作。

> 在這個計畫裡同樣重要的還有身體感，不同年齡、職業、地區的人，會有不同的身體感，而這些身體感都能夠成為系列工作坊的著力點。探監搭火車時的身體感、軍事勞動的身體感、躲在甘蔗田中的身體感等，雖然今日的境況不完全相同，但身體感受仍會有能共通之處。這些身體感得以成為民眾「共感」的橋樑，讓參與者在身體工作坊時更能夠發揮，身體工作坊時拍攝下的畫面也就意味著民眾對這段歷史的反應。創作出畫面關係是身體工作坊的用意所在。（文｜蔡喻安）

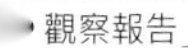

觀察報告

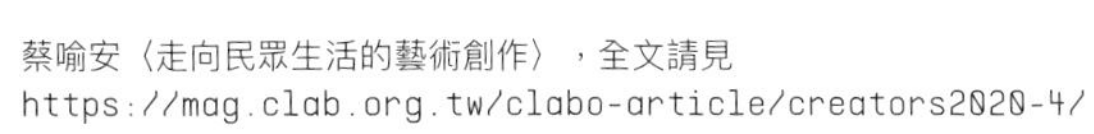

蔡喻安〈走向民眾生活的藝術創作〉，全文請見
https://mag.clab.org.tw/clabo-article/creators2020-4/

神的棲所 GiR
God in Residency

在寫實表演中「角色進到演員」的說法與「神明降駕乩童」在邏輯與概念上似乎有著高度的關聯性。同時這兩種說法也暗示了心與腦二分的預設，彷彿演員與乩童的肉身是一特殊的媒介，得以透過特定方法邀請不同的心靈進駐，也同時產生不同的「身分」。然而，在「進駐」的當下，「自我」會在哪？它又以什麼形式存在？一個肉身裡是否能有複數的「自我」？若「表演」作為人類社會中可系統化、可傳授的「技術與知識」，是否表示「起乩」也有機會是可以系統化的「技術與知識」？若我們能夠透過對於表演技術與實踐試圖建立起乩的技術與方法論，或許不久後就能透過教育讓每個人都直接地「與神共存」，讓神權不僅僅只是特定人被賦予（被擇選）的特殊權利與資源，而是人人都有機會習得的技術。「神的棲所」比喻神降駕於人體時，乩童便是神的短期居所，透過表演理論與科學實證的交互實驗下，我們是否有機會能「解放神權」並再次趨近「眾神在世」，邁向「神／人」間「跨物種」的民主世界？本計畫試圖交匯與連結「腦與神經科學」、「乩童文化與展演」、「表演方法與技術」間相互驗證與探索，尋找能創造出更多「神的棲所」的可能方法。

There seems to be an intimate logical and conceptual association between “a role occupying its actor” in realistic performance and “a deity descending on a spirit medium” in religious belief. The two situations also suggest the mind-brain dichotomy, as if the body is a special medium that employs specific ways to invite different minds to enter it, hence different “identities.” However, where is the “self” and what form does it take at the moment when a mind “enters” the body? Can the body accommodate plural selves? If “performance” as “expertise” is systematic and teachable, does it imply that “deity possession” belongs to such kind of “expertise” as well? If we can develop the technique and methodology of deity possession through performance, maybe we will be able to allow everyone to directly “coexist with deities” through education in the near future, and turn the divine right from the privilege and resource exclusive to a specific group of people into the technique everyone can learn by being taught. The metaphor of “God in residency” is used to represent that a spirit medium serves as the short-term residence for a deity when the latter descends on the former. Is it possible for us to “liberate the divine right,” “co-exist with deities,” and create a democratic world for the hybrid of “deity/humanity” by applying performance theories and empirical research? This

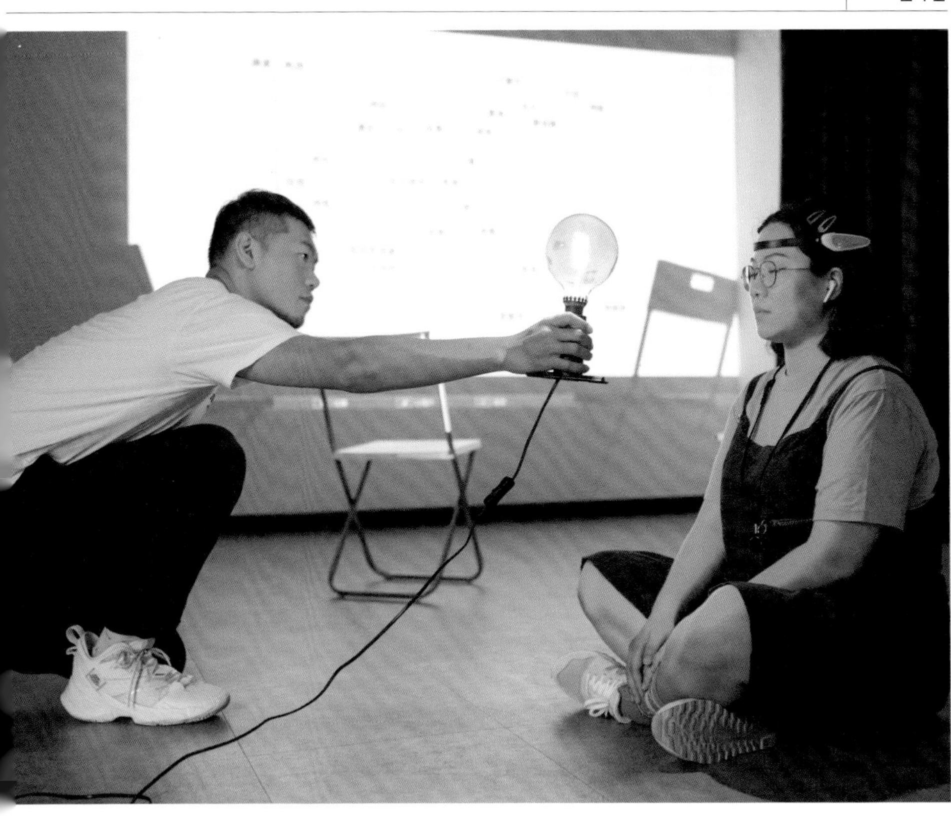

project attempts to explore the connections among "human brain and neuro-science," "spirit medium culture and performance," as well as "performance approaches and techniques," thereby discovering more ways to achieve "God in residency" within the realm of possibility.

REATORS

鼎云 HUANG Ding-Yun

術家、劇場編導。目前為「明日和合製作」共同創作。作品包含共同創作、空間回與跨領域實踐，專注調動觀眾與表演者的觀演關係的框架。近年作品共同創作分：2019 年臺灣國際藝術節《半仙》、18 年臺灣當代文化實驗場「再基地：實驗成為態度」之《林愛國計畫》等。

HUANG Ding-Yun is an artist, theater writer, and director. He is one of the co-founders of Taipei-based multi-creator collective, Co-Coism. Co-coism aims at work-in-collective, site-responding and interdisciplinary practices. They focus on creating a flexible relationship between the audience and the performers. Co-coism's recent works are *Play God* at 2019 Taiwan International Festival of Art, *Where is Ai-Guo Lin?*, presented in Re-Base: When Experiments Become Attitude, at C-LAB in 2018, etc.

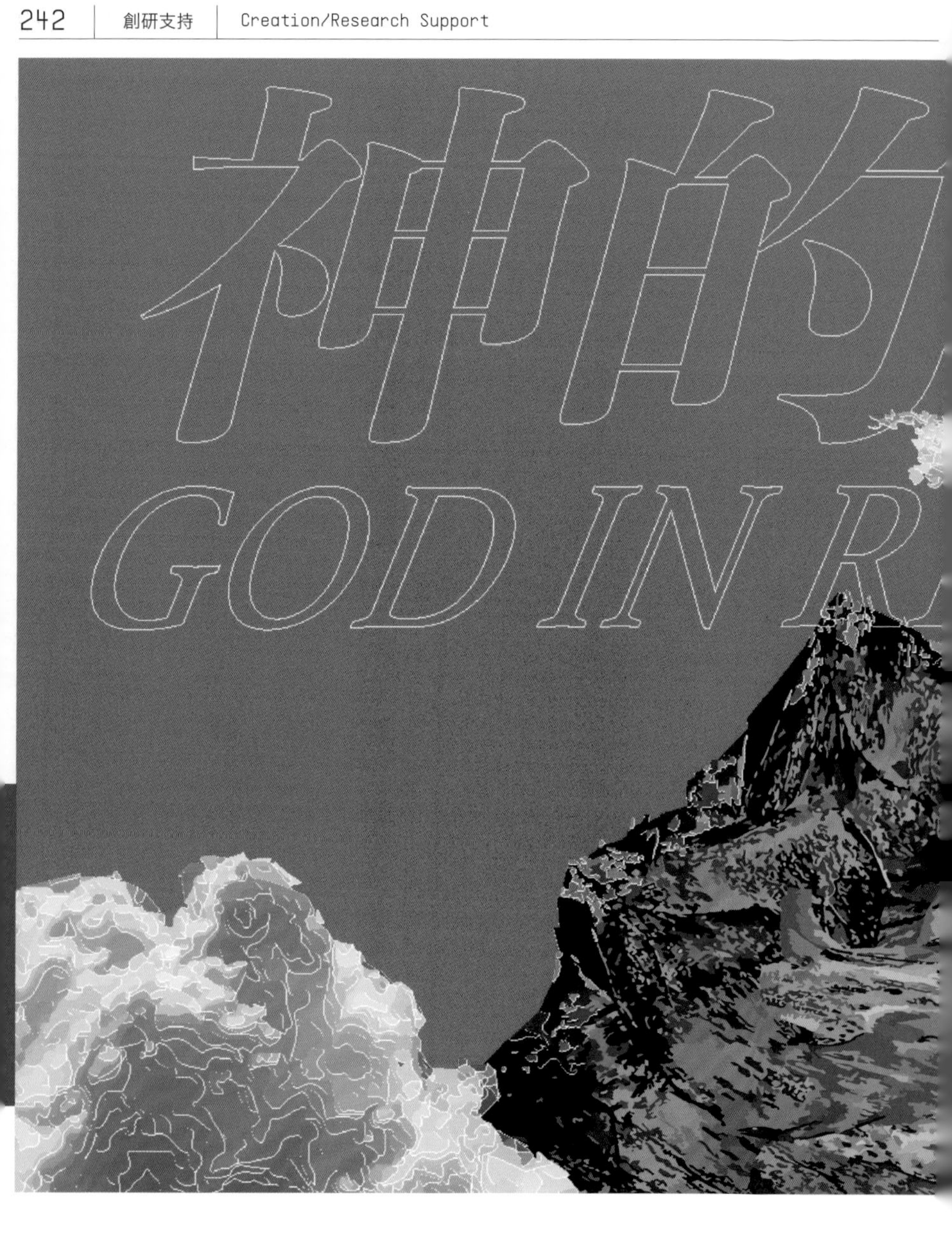
神的
GOD IN R

▌黃鼎云，「神的棲所 GiR」，2020_

HUANG Ding-Yun, *God in Residency*, 2020_

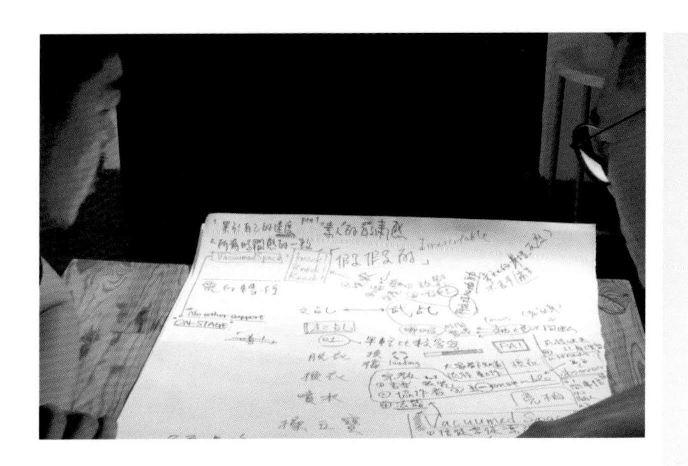

操演 再現

角

自我

神

> 藝術創作／信仰（精神狀態）為何需要客觀數據的驗證？黃鼎云在該計畫中，試圖透過腦波儀偵測器來探究神降乩身與演員上戲狀態之間的差異。如果情緒可以透過實證科學測量的數據得到情感的量化圖表，以作為理解人類情緒的方法。屬於超自然的、想像的情感波動，是否也能有另一種不是透過展演而是數據的理解方式？藝術家的文化實驗即是在心靈與腦之間，這樣的藝術感性能否被科學理性驗證？

> 明日和合製作所在 2019 年的臺灣國際藝術節中的演出《半仙》，持續將真實人物置放到劇場空間，前半段以表演講座（lecture performance）形式呈現，三位主創者作為表演者，透過問事遂展開赴上海的「找尋前世」之旅，論述與述敘並行，最後宮主「起乩」（濟公）現場問事約十五分鐘，在此劇場空間又進行了虛構展演與現實問事的曖昧交疊，將現實情境置放到劇場空間，在表演空間中進行巫宗教的日常演示，所謂表演中的表演，究竟只是講座形式的劇場表演？抑或是展示何謂日常巫宗教儀式的表演，問事的際社會功能在劇場空間是否被消解？從此刻的劇場中的「政治」詮釋角度回望，明日合製作所已經在探問所謂政治治理與表演政治的關係。

> 或許，對黃鼎云來說，真正值得懷疑的，是神的預言究竟是怎麼發聲及傳達？此次他的計畫以演員演乩童、以儀器作為中介，試圖建構一套人人都可以成為乩童／自我操作者，為自己的生命難題解惑進行自我通靈的實驗。如果每個人都能成為一位能自我通靈與解籤的乩童／問世者，或許就能解除他者轉譯神諭的這層關係？真正的問題或許不是所有人都能真的學習一套降駕術來自我圓滿，而是們如何抵抗現實社會充滿代替我們發言的權力者位置。當政治人物操弄信仰的時候，政治人物成為的話語的絕對詮釋者，詮釋將神性變成政治籌碼的時候，人民應該起身學習成為乩童的方法，以直獲得神諭。我們可以說，黃鼎云在乎的不是個體的信仰與其內容，而是信仰的被政治化。（文｜羅倩

觀察報告

羅倩〈施懿珊的資訊（意識）戰與黃鼎云的預言（治理）術〉，全文請見
https://mag.clab.org.tw/clabo-article/creators2020-2

精神與靈魂的治理之術
The Technique of Mind and Spirit Control

訊息結構是一層一層的「裏空間」構成，我們以為的「裏部」有時候只是另一個裏部的「表象」。「精神與靈魂的治理之術」計畫，借用研究病毒傳染的經典模型——SEIR，作為對「敘事」的研究路徑，來解構臺灣與中國兩地訊息的「關係網絡」。並依照計畫研究得到一個基於事件的「敘事（標籤）模型」，以此模型作為「視覺框架」，進行（數位）虛構空間的敘事建構活動。過往數位內部空間的結構，都是靠對現實的再現，但「精神與靈魂的治理之術」要處理的是一個由網路社會「標籤」生成的世界結構。而這個運算原理製作出的世界結構，就形成一種新的地景、地緣。

The structure of information comprises layers of "inner space." The "interior" we think is sometimes the appearance of another interior. This project uses the well-known epidemiological model SEIR as the research approach to "narrative," seeking to deconstruct the cross-strait "relational network" of information. This project develops an event-based "narrative (epithet) model" as its "visual framework" for the narrative construction in a (digital) virtual space. The structure of a digital space used to rely on the representation of realities, while this project is intended to tackle the world structure generated by "epithets" in the network society, a structure that constitutes a novel landscape, a new geographical context.

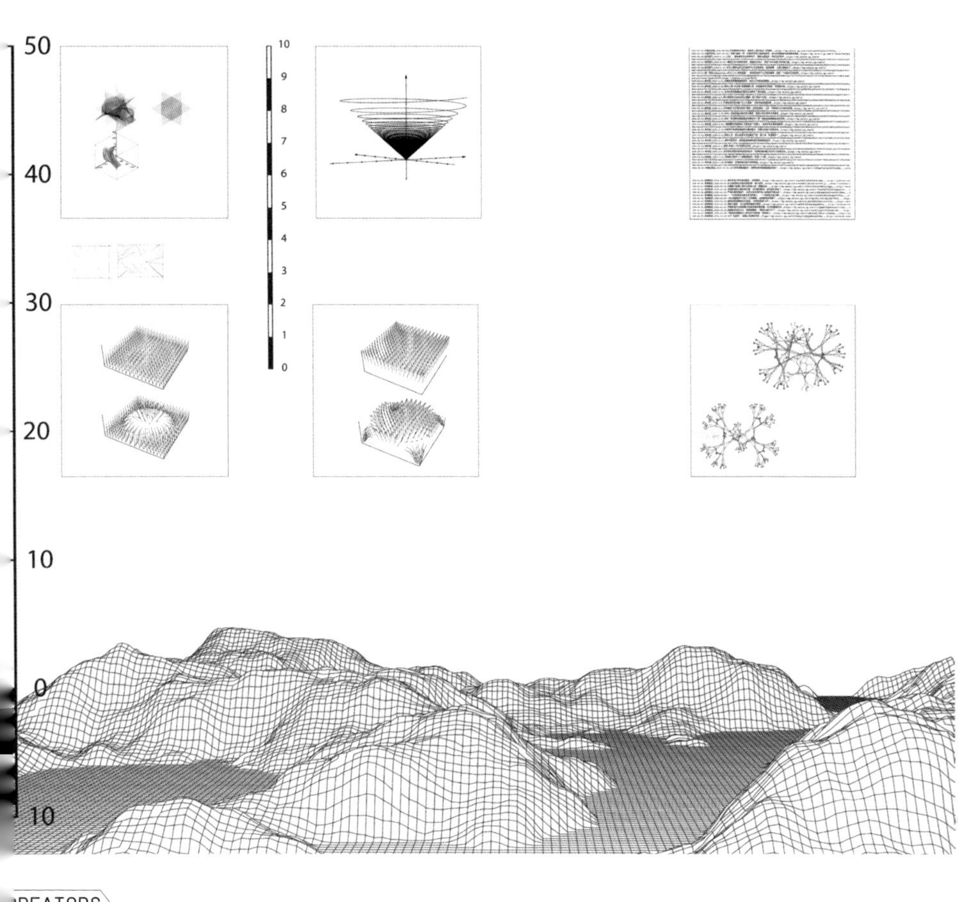

CREATORS

施懿珊 SHIH Yi-Shan

空中自體動力宣言（創辦人）。自 2019 年起開始常態參與由虛構研究者——賴火旺發起的「議題串連」。當代的集合場域，其實是由各種議題標籤組成，就像 hashtag 的觀念。你對這個議題認同或關注，你就會在這個議題框架下被連結在一起，當下集合場所也立刻成形。而議題解散了，相對地、群體和空間就解散了——這就是「議題串連」截至目前為止的運作模式。

SHIH Yi-Shan is the founder of Body Aircraft. In 2019, she started to participate in Issue Connection initiate by a fictional researcher LAI Ho-Wan. The structure of contemporary gathering platforms comprises of multiple issue tags, which are similar to hashtags conceptually speaking. People would become mutually connected if they recognize or pay attention to the same issues, gathering spaces are born simultaneously. Groups and spaces would be dissolved the moment specific issues are cancelled. This is how Issue Connection operates for now.

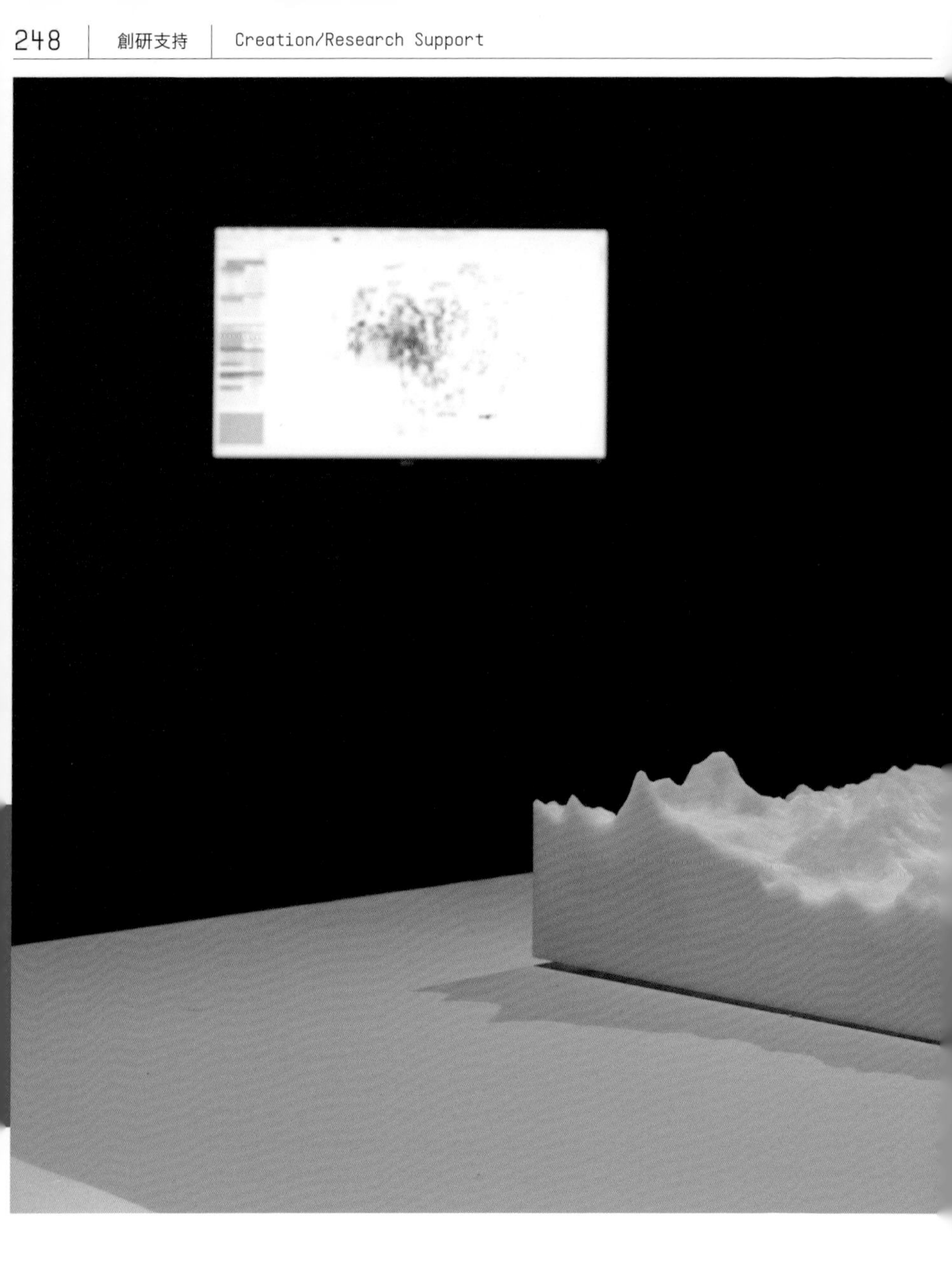

▌施懿珊，「精神與靈魂的治理之術」，2020_
SHIH Yi-Shan, *The Technique of Mind and Spirit Control*, 2020_

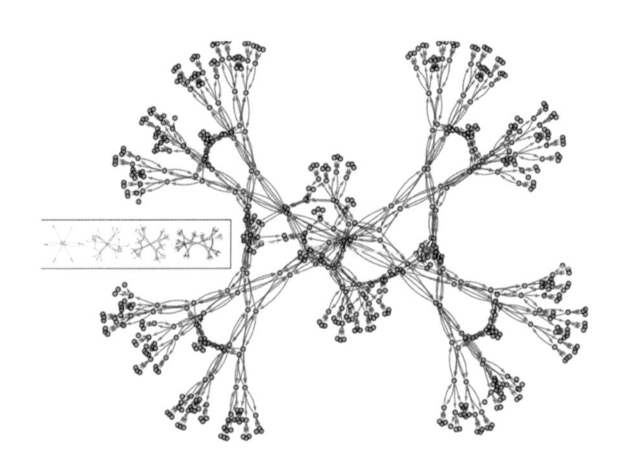

> 與其說施懿珊是透過找尋網路世界中的敘事法則（在提案計畫的目標是武漢肺炎疫情敘事鏈）進行藝術創作，不如說她是透過尋找敘事法則的模型，探究模型背後究竟是什麼樣的「意識」在控制人的思想與生活。也就是目前最顯而易見，卻同時難以辨別的網路資訊戰（爭）。例如：意識形態偏頗的報導與試圖在短時間混淆視聽的假新聞。經常性地影響大眾對事件的看法與判斷，意識形態的操盤手透過資訊戰藉以改變民意風向，進而影響選舉或重大決策。

> 以她在中國網域持續多年的網路田野，經常在公開講座分享在中國網路的實際經驗：政權如何對人民在網路上的言論進行管控。例如她因為誤觸「蛤蟆」關鍵字，使得經營多年的神秘怪物論壇被消失。因為「蛤蟆」一開始其實是中國網民對政治人物江澤明言行的網路迷因（meme），後形成「膜蛤文化」（也稱作江蛙崇拜），「蛤蟆」因為影射了對政治人物的批判，最終成為在特定網路上被言論管制（消失）的關鍵字。

> 對施懿珊來說，她的核心關懷是找出當代人生活在何種社會情境之中。如果我們要確實知道自己 於什麼樣的當代世界，破除魔障的方式，是必須清楚地認識我們時時刻刻面對到的是何種樣態的 事型，這左右我們的思想與生活，敘述邏輯如何影響決策當下與判斷未來。換言之，藝術家在找 的是當代世界運作的敘事型，最切身的是我們與中國、美國的政治及經濟關係，並透過藝術創作 造一個敘事模型，讓人體驗到我們究竟在什麼樣的敘事環境裡生活。

> 從現實中的「軟抗爭」與博物館式展演的關係，到網路分身與美術館機構的關係， 懿珊依然在面對一位藝術創作者與美術館展覽機制的角力，做藝術如何觸碰現實社 的真實樣貌，並同樣能透過展示帶來力量，而不是由美術館的展覽機制抵銷藝術的 正潛能。（文｜羅倩）

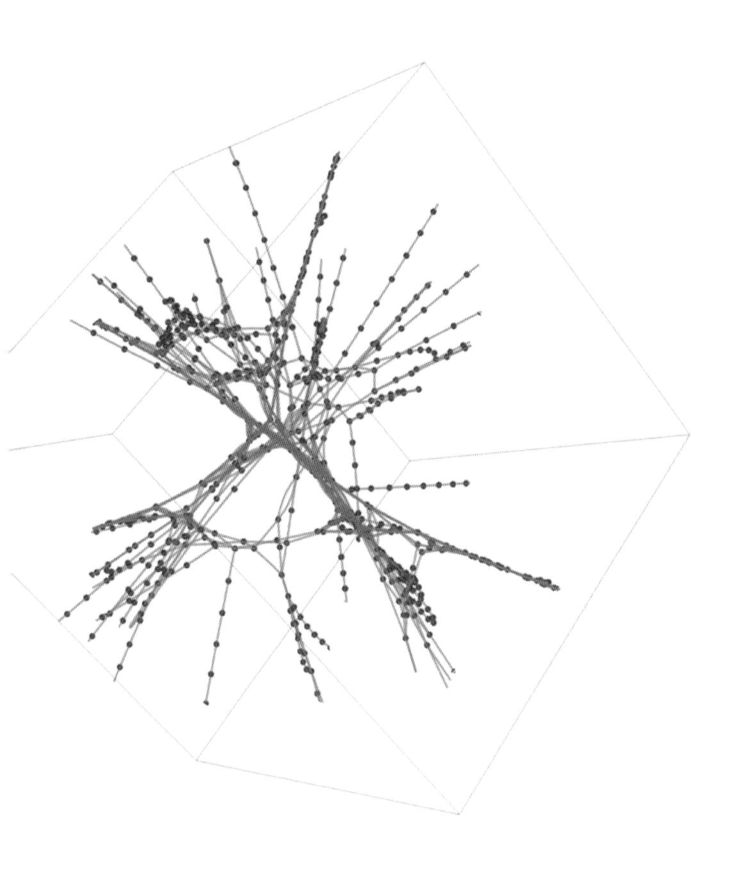

觀察報告

羅倩〈施懿珊的資訊（意識）戰與黃鼎云的預言（治理）術〉，全文請見
https://mag.clab.org.tw/clabo-article/creators2020-2

七個在海上的人：賽鴿
Seven People Crossing the Sea

《七個在海上的人：賽鴿》是一個錄像作品，受香港布料市場（棚仔）攤主——何應的生命故事和他發起的反棚仔迫遷運動所啟發，以何應的故事為主軸並邀請他作為演員參與拍攝。「棚仔」自 1978 年靠著遷入的攤商自我組織、經營，逐漸有機地發展成一個有系統的布料市場。自 1960 年代起，香港作為歐美拓展全球成衣產業生產鏈的中繼站，眾多港商因語言、地利優勢，紛紛在亞洲各國投資設立紡織廠與成衣加工廠，利用當地相對廉價的勞動力，向當地外包、分包來自歐美的龐大訂單，而棚仔布販之貨源，往往來自於亞洲具港資背景的廠商所用剩的零星布料，俗稱「士多布」（stock 布）。待售布料層層堆疊，橫跨 40 年，香港與亞洲成衣產業的盛衰，都被壓縮其中。何應是廣東佛山人，年輕時與朋友一行七人，泳渡大海偷渡進入香港。他目前是「棚仔」內碩果僅存的攤主之一，也是「棚仔」反迫遷運動的發起者。棚仔面臨被香港政府迫遷的問題，引起了小販的合法性／正當性、基層生活空間等的探索和思考。何應所發起與現存布販組織組成「棚仔關注組」，並與學生、老師、社會運動者、記者、設計師、藝術家合作，思考並發掘棚仔存廢之外，在歷史上、文化上、時代上所代表的意義，試圖創造多樣的價值和可能性。

2020

This project is a work of video art inspired by the life story of HO Ying, a stall owner in the Pang Jai Fabric Bazaar, Hong Kong, and the anti-relocation movement he started. HO is invited to join the cast of this video that features his story. Through the self-organization and management of the stall owners who moved to this place since 1978, it has evolved organically into a systematic fabric market. Since the 1960s, Hong Kong has been the relay station for occidental countries to expand their global garment industry chains. Taking advantage of their language proficiency and Hong Kong's geographical location, many Hong Kong businessmen began to set up textile factories and garment processing factories in Asian countries, in order to harness the relatively cheap labor there. They tended to outsource and subcontract the orders from occidental countries. In the Pang Jai fabric bazaar, the cloth sold by the vendors, commonly known as "stock cloth," usually came from the cloth scrap left by Hong Kong-funded manufacturers. Different types of cloth were piled up, encapsulating the 40-year vicissitudes of the garment industry in Hong Kong and Asia. As a native of Foshan, Guangdong Province, HO Ying sneaked into Hong Kong by swimming across the sea with his six friends when he was young. He is one of the few remaining stall owners in the bazaar as

well as the initiator of the anti-relocation movement. Forced to be relocated by the Hong Kong government, the Pang Jai Fabric Bazaar has stimulated the reflection on the vendors' legality/legitimacy and the exploration of the grassroots living space. Initiated by HO Ying and organized by the remaining cloth vendors, the Pang Jai Concern Group collaborates with students, teachers, social activists, journalists, designers and artists, striving to contemplate and discover the bazaar's historical, cultural and contemporary significance beyond its preservation and demolition, insofar as to offer it added value and diverse possibilities.

REATORS

博志 HUANG Po-Chih

博志的藝術實踐著重探討的是農業、造業、生產、消費等議題，皆與其家處境和家族歷史相關。2013 年出版散集《藍色皮膚：老媽的故事》，透過錄他的母親，一個平凡個體的職業流，去間接反映臺灣近 50 年來的社會改與經濟變遷。他近年的創作計畫《五百檸檬樹》，將展覽作為募資平台，一面挪用藝術世界的資源發展農業品牌、化廢耕地、種植檸檬樹並釀造檸檬酒；一方面聯繫家族成員、當地農夫、消者，產生新的社群關係。

HUANG Po-Chih's diverse artistic practice revolves around the circumstances and history of his family which enable him to involve in issues like agriculture, manufacturing, production, consumption, etc. In 2013, he published his first collection of essays *The Blue Skin: My Mother's Story*. By documenting the story about his mother, the career change of a normal person, the book reflects Taiwan's social reformation and changes over the past fifty years. His recent project *Five Hundred Lemon Trees* has been transformed to a crowd funding platform allowing the appropriation of artistic resources for developing an agricultural brand, activating fallow farmland, and growing lemon trees for lemon liquor. On the other hand, the project has connected his family members, local farmers and consumers to make a new social relationship possible.

他總認爲要走的時候，
走就對了

黃博志，「七個在海上的人：賽鴿」，2019-2021_

HUANG Po-Chih, *Seven People Crossing the Sea (work documents)*, 2019-2021_

› 黃博志所撰寫的〈酸澀草樹〉是何應青少年時期的故事，在故事中，他與朋友一行七人，計畫從中國廣東偷渡到香港，為了追尋屬於他們的香港夢。他們每天勤練游泳六小時，閒暇之餘就打零工籌錢，自建舢舨船。一年過後，他們一身輕便地扛起舢舨船，沿著西江而下，從澳門出海。出海不久船身便不敵大浪翻覆，他目睹體力不支力竭而亡的同伴，他說自己也死了好幾次，「一次、兩次、三次……」。最終，他幸運地爬上一座小沙洲島，島上僅有的一棵樹，是貌似在家鄉俗名「酸澀草樹」的植物，他爬上樹，奮力啃食樹上每一片樹葉和果實，隨後睡意侵略他的意識，也偷走夢境，失去時間感的沉睡了，他說：「舒服！」緊接著一天一夜泳渡珠江口，登陸大嶼山，他再次沉睡，但這次，他奪回了他的香港夢。

› 黃博志的創作，長期以龐大的消費體系為背景，分析複雜的產業結構，並深入結構之中的個體，尤其這其中長期被忽略的勞動群體。他改編那些以紡織、製衣業維生的人們的故事，訪談關於他們的「夢」——那些在時代推進及產業演變下被遺忘的生命經驗，各自所懷抱的現實、虛構、夢境般地不的夢。某種程度下，這些夢都述說著時代潮流下，對於現實的不同想像。而黃博志將訪者的「夢」、相處當下的對話以及想像書寫成故事，並邀請受訪者參與演出拍攝成像作品。但拍攝沒有固定腳本，強調現場直覺感受。

› 近十年的創作，不單純是議題的討論，或是旁觀地述說他人的生命故事，自己也常轉換不同的身並身陷其中。黃博志的創作和生命有著緊密難分的狀態，彷彿生命本身就是創作，每個創作計畫此牽動，甚至無法有明確的開始與結束。因此，生命中無法預測的變數，例如那些突如其來的病死亡，都影響著他的創作狀態與內容。（文｜王瑀）

觀察報告

王瑀〈臺灣藝術創作計畫，先有雞？還是先有蛋？有雞就一定有蛋嗎？〉，全文請見
https://mag.clab.org.tw/clabo-article/creators2020-5/

綠頭鴨之死
The Unusual Death of a Mallard

計畫靈感源自一隻綠頭鴨，牠是鹿特丹自然史博物館的藏品。1995 年在博物館新建的玻璃場館外，一隻雄性綠頭鴨撞死在玻璃牆上，意外發生之後，引來了另外一隻雄性野鴨。活著的鴨強暴了死去的鴨，過程長達 75 分鐘。博物館館長 Kees MOELIKER 目擊了整個事件經過，將鴨子製成標本、於博物館展示，並寫了一篇學術報告〈首例野鴨的同性戀屍癖行為〉；此份報告贏得 2003 年「搞笑諾貝爾生物獎」。在這個計畫中，我們將探索博物館典藏技術是如何展示動物的身體／記憶。從博物館所使用的工具出發，如標本製作、數位典藏的 3D 掃描、生物研究熱感應成像等，檢視工具自身的倫理問題。

This project owes its inspiration to a mallard, a collection item at the Natuurhistorisch Museum Rotterdam. In 1995, a male mallard died of hitting the glass wall of the newly built glazed pavilion of the museum. This accident attracted another male wild duck, and it “raped” the corpse of the dead mallard. This event lasted 75 minutes. Witnessing the entire process, the museum director Kees MOELIKER turned the dead mallard into a specimen and exhibited it in the museum. He also wrote an academic report titled “The first case of homosexual necrophilia in the mallard Anas platyrhynchos,” which earned him the 2003 Ig Nobel Prize in Biology. This project is intended to explore how museums use technology to display animals’ bodies/memories and examine the ethical issues concerning the tools adopted by museums such as specimen production, 3D scanning in digital archiving, and thermal imaging for biology.

REATORS

哲瑜 HSU Che-Yu

術創作者，畢業於國立臺南藝術大學
形藝術研究所。現參與HISK（比利
高等藝術學院）為期兩年的進駐計畫
ost-academic residency）。

An art practitioner and a graduate of the Graduate Institute of Plastic Arts, Tainan National University of the Arts, HSU Che-Yu (b. 1985) is undertaking a two-year post academic residency program at the Higher Institute for Fine Arts (HISK).

琬尹 CHEN Wan-Yin

作者，畢業於國立臺南藝術大學動畫藝
與影像美學研究所。現就讀於阿姆斯特
自由大學（Vrije Universiteit
sterdam）藝術史與文化學系博士班。

A writer and a graduate of the Graduate Institute of Animation and Film Art, Tainan National University of the Arts, CHEN Wan-Yin (b. 1988) is enrolled in the Ph.D. program of Art History and Cultural Studies, Vrije Universiteit Amsterdam.

HIGHWAY

▌許哲瑜和陳琬尹，「綠頭鴨之死」，2020-2021_

HSU Che-Yu and CHEN Wan-Yin, *The Unusual Death of a Mallard*, 2020-2021_

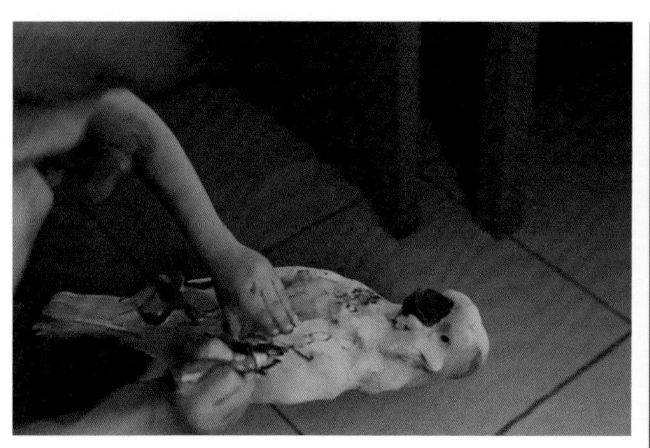

> 「綠頭鴨之死」以多種技術記錄不同狀態下的學術用動物，混製成為錄像作品。真偽觀影的觀賞體驗，也延續許哲瑜和陳琬尹兩人一直以來對人類記憶中真實與虛構議題的關注。

> 「跨領域」發生時，雙方是平等交流，還是必然會有主從之分？當藝術家將生科領域的專業作為跨領域創作的內容時，是從何種角度出發去進行交流？跨界是很互相的事情，同時也是基於不同領域背景的人該如何互相理解的溝通過程。在生物研究這行裡，研究者須不斷產出符合圈內價值觀的作品，才能累積履歷進而生存下來。藝術領域也有類似的狀況。蜻蜓點水式的跨域實踐，雖然常挪用或誤用別人的知識，但卻能在不挑戰本業價值觀的前提下，為自己的經歷注入活水，反而是務實可行的方案。（文｜王順德）

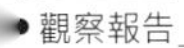

觀察報告

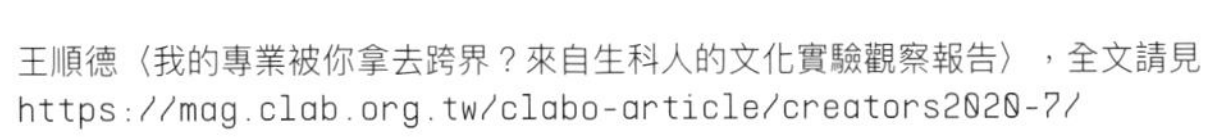

王順德〈我的專業被你拿去跨界？來自生科人的文化實驗觀察報告〉，全文請見
https://mag.clab.org.tw/clabo-article/creators2020-7/

負地理學：近未來的蓋亞實踐與藝術計畫
Neg-geography：Toward Practice of Geo-art

「負地理學」是關乎一項地理實踐與負熵預想的試驗場域，探問在地的科幻敘事和地心想像，以及島嶼的地質、地熱、地層、火山、斷層帶、板塊交界帶，這些物質基礎如何讓藝術實踐有機會與地質學、考現學、媒介生態學、歷史學、自然地理學的知識彼此交織與對話。甚至將地質作為一種文化的基底，積極思考它與文化實踐的關係，以多樣性的地理知識，進行某種感知學、人類存有與事物價值的重估。因此，我們需要發展一個去探索地理、地質如何影響感知、藝術與創作界限的工作隊或調查隊，在幾近半年的時間裡，藉由火山潛勢線與地熱區、斷層線與城市考掘、非關災難的紀錄劇場、跨域座談、影像共創、適地書寫等方式，從各種差異的敘事介面，集結出一種行動網絡，實踐走在土地之下的根系視野——轉換視角、編織知識、重返大地的行動方案。從而思考臺北作為「火山城市」，人們如何棲居於大屯火山組構的地理生態群落之中。

"Neg-geography" is tantamount to a test site for geographical practice and negentropy presumption. In addition to examining local sci-fi narratives and imagination about the earth's core, it addresses the question as to how the island's material foundation (e.g. its geography, geotherm, strata, volcanoes, fault zones, and plate subduction boundaries) engages artistic practice in a close dialogue with geology, modernology, media ecology, historiography, and natural geography. Furthermore, it treats geology as a sort of cultural heritage and tries to clarify its relation to cultural practice, so as to cognitively reassess the raisons d'être of human beings and things with a rich body of geographical knowledge. Hence we need to build an expedition team to explore how geography and geology push the limits of our perception and artistic creation. During this period of nearly six months, we will weave a network of actions through different narrative interfaces and by reference to volcanic eruption potential lines, geothermal zones, fault lines, urban archeology, non-disaster-related documentary theaters, transdisciplinary talks, image co-creation, and location-based writing, thereby opening up rhizomized horizons – an action plan that switches our perspective, advances our knowledge, and brings us back to earth. By doing so, we can

contemplate how the residents in Taipei (as a “volcanic city”) dwell in the geographical ecology constituted by the Tatun Volcanoes.

REATORS

引爆火山工程
ngineering of Vocano etonating

016 年，梁廷毓、許博彥、盧均展、盧冠宏組成的團隊開啟了《引爆火山藝術行動》，在陽明山地區進行了一系列的燃煙計畫之後，在 2017 年進入城市與火山的區域關係探討的創作，透過對 2010 年代新聞媒體報導的的火山形象，探討火山如何透過媒體被恐怖化，以及城市將自然他者化的問題。

The artist group formed by LIANG Ting-Yu, HSU Po-Yen, LU Chun-Chan, and LU Guan-Hong launched the *Engineering of Volcano Detonating* in 2016. After carrying out a series of smoke rising projects in the Yangmingshan National Park, they engaged in the artistic creation concerning the relations between the city and volcanoes in 2017. Applying the image of volcanoes in the media coverage throughout the 2010s, they investigated the demonization of volcanoes by mass media and the othering of nature by the city.

▍引爆火山工程，「負地理學：近未來的蓋亞實踐與藝術計畫」，2020_

Engineering of Volcano Detonating, *Neg-geography: Toward Practice of Geo-art*, 2020_

> 此次引爆火山工程的計畫以相對複雜的形式呈現：包含了論文發表、系列講座、火山區域的實地探訪（「夜行者計畫」）、視覺呈現與預計的出版計畫。如此複雜的形式或許可以追溯到團隊成員梁廷毓於 2019 年 11 月的個展「墳 · 屍骨 · 紅壤層」，此次個展同時也以相對複雜的形式呈現：兩個展區、展覽對談、系列講座、工作坊以及於人文學社年會的論文發表。關於為何以如此複雜的形式進行創作，或許不能單純地以「計畫型創作」一詞帶過，而有必要加以深究。

> 回顧此次個展的開幕對談，對談人王聖閎從雙年展在臺灣的近況談起，他認為自「2012 台北雙年展：現代怪獸／想像的死而復生」起，無論是聚焦在歷史、人類世或是檔案等問題，都建立在相同的預設之上：知識的可見性（視覺藝術中隱含的可視覺化邏輯）以及可立即交換。他並且指出，越學術性的展覽便越傾向於將知識的可交換性作為前提。但某些不可見的、時間上不允許快速流通與交換的知識，該如何避開這些限制呈現？更具體地說，為了對抗這些當代視覺藝術中的預設，必須採用何種方式來呈現如長期的、持續性的田野經驗，或是某些外於我們認知框架的「地方知識」？藝術家嘗試給出的回應便是我們看到的複雜形式，也就是各種不同的展示介面。（文｜沈克諭）

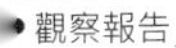

觀察報告

沈克諭〈不在此時：時差的潛勢〉，全文請見
https://mag.clab.org.tw/clabo-article/creators2020-1/

糖蜜、酒精、健身工坊，是什麼使今日的生活變得如此不同，如此有魅力？

Molasses, Ethanol, Fitness Workshops, Just What Is It That Makes Today's Life So Different, So Appealing?

蔗糖從來不是一個必需品，但它卻改變了世界的味蕾並使人成癮。昔日，人類為了追求蔗糖展開了無數的貿易、征戰與殖民；今日，人們為了雕塑身形而開始戒斷糖份與積極健身。人類對糖的需求，在不同的時代脈絡下展現了多重的定位。蔗糖的價值與定義，從農產品轉變為經濟貿易物資，並逐漸形成社會中難以戒斷的文明徵候；蔗糖產業也在不同的時代脈絡下扮演不同的社會功能，甚至透過自由貿易的擴散，促使熱帶商品經濟在殖民地產生反轉的動能。本計畫欲透過蔗糖產業所展開的歷史脈絡與想像空間，探討熱帶經濟作物在全球貿易體系中的歷史與權力、貿易與流動、現代化與加工，以及在自由貿易下的殖民地反向流動，並透過藝術實踐來回應在全球化所開啟的生活情境，思考人類對美好生活的想像與塑造，呈現蔗糖產業在不同時代中的角色定位與價值轉換。

Sucrose has never been a necessity of life. However, it's so addictive that it altered the gustation of humankind. Previously, human beings engaged in trade, wars and colonization for sucrose. Nowadays, people start to abstain from sugar and work out actively in order to keep fit and healthy. Accordingly, sugar has played different roles in different historical contexts. Sucrose's value and definition evolved from an agricultural product into a tradable commodity. Today, it caused a civilized social symptom from which escape is nowhere on the horizon. The sucrose industry has performed different social functions in different periods of time as well. The proliferation via free trade even boosted the tropical commodity economy in colonies. By virtue of the historical development and imagination unfolded from the sucrose industry, this project seeks to investigate the history, power, trade, circulation, modernization and processing of tropical cash crops as well as the reversed flow in colonies under the free trade system. This project also responds to the social circumstances with artistic practice in the era

of globalization, contemplating people's pursuit of a better life and presenting the varied roles and value of the sucrose industry in different periods of time.

REATORS

羅懿君 LO Yi-Chun

懿君作品以自然媒材為素材，關注臺灣經濟作物的貿易歷史脈絡，自 2013 年迄今，陸續採擷香蕉皮、菸草、甘蔗渣為媒材，創作出不同形式的物件與空間裝置，透過田野調查、資料蒐集與文獻閱讀，耙梳臺灣產業地景的歷史脈絡與世界經濟體系座標。探索臺灣自海權時代以來，作為串連東亞與世界經濟鎖鏈中的樞紐位置，到全球化的世界裡，人與物的流徙。

LO Yi-Chun is interested in the trade history of Taiwanese cash crops and adopts natural materials as her creative media. Since 2013, she has not only created objects and spatial installations in different forms with banana peels, tobacco leaves and bagasse, but also collated both the historical context of Taiwan's industrial development and the structure of the global economic system through field survey, data collection, and literature review. She tries to examine Taiwan's pivotal position that has connected East Asia with global supply chains since the Age of Discovery, as well as the flow of people and goods in the globalized world.

羅懿君，「糖蜜、酒精、健身工坊，是什麼使今日的生活變得如此不同，如此有魅力？」，2020_

LO Yi-Chun, *Molasses, Ethanol, Fitness Workshops, Just What Is It That Makes Today's Life So Different, So Appealing?*, 2020_

> 糖蜜是第一個關鍵字，也是整個創作的基礎核心。羅懿君過去便曾以菸草葉、檳榔等作物作為創作素材，2019 年時參與了「麻豆糖業大地藝術祭」，開啟以糖為主題的創作計畫。蔗糖製造過程中的糖蜜，能被發酵成為酒精，糖蜜酒精是日治時期日軍使用的燃料之一，也是這個計畫的第二個關鍵字。糖是民生物資，但它變成糖蜜酒精時就成為了軍事物資。羅懿君當時便已經以這個轉化關係作為創作主題，這次又往前推進了一些，他把現代社會的糖也放入創作當中，也就是這個計畫的第三個關鍵字：「健身」。乍看之下是一個很晦澀的創作計畫，但其實不然，計畫以一種歡快趣味的姿態進行著，就如同許多人原本對「健身」這件事的印象一樣，你想像他是乏味，但當你開始做了之後就會發現事情並不是你以為的那樣，其實更有趣的多。

> 除了工作坊外，羅懿君各拍攝了一支自己訪談健身教練與開甜點店的朋友的影片，訪談內容不外乎糖與健身。選用影片拍攝這個形式亦有其意義，對羅懿君來說，他前往健身房時常常看到的就是民眾邊做運動，邊看著健身房中電視螢幕的畫面這一幕成為他開始用影像拍攝與呈現的靈感。在這項計畫裡，甘蔗與健身以各種既跳又合理的方式連結了起來，可以說他們分別意味了糖的傳統與糖的現在。這個今與昔間的並存，是這整個計畫一以貫之的風格，可以說也是這個計畫最有趣的部分。羅懿提起甜點店影片的拍攝，便是再三強調引起他興趣的正是健身與甜點兩者之間乍看之的違和感。健身這個元素在這個產業計畫裡看似突兀，但想到深處，卻又是無比自然關乎健身的部分，其實都是在對現在的人們提問：「對（現在這個時空的）你而言，是什麼呢？」（文｜蔡喻安）

觀察報告

蔡喻安〈走向民眾生活的藝術創作〉，全文請見
https://mag.clab.org.tw/clabo-article/creators2020-4/

凝結歷史殘響：臺灣洞穴的感知敘事
Freeze Historical Reverb: the Perceptual Narratives in the Caves in Taiwan

因為特殊的海島地形與豐富而險峻的發展歷史，臺灣有許多洞穴——防空洞或是隧道。這些洞穴多半是人工的（為了軍事或者交通運輸的目的），少數是自然形成的。這些洞穴的故事透過文字系統進入我們的知識裡，在這樣的過程中我們的感官(sensibility）都不需要作用。現代化中的我們，感知（perception）和知識往往是脫離的。

本計畫實驗翻轉這些「知」和「感」的關聯與順序，採集臺灣各處不同的洞穴故事，利用天然的特殊聲場，以聲音演繹這些故事中第一人稱的聽覺經驗，並以Ambisonic 及雙耳立體聲（binaural）編碼格式錄音重現，創造出一段段沉浸式的聲音旅程。沉浸式的聲音體驗讓參與者以第一人稱的視角，啟動自己的感知，形成對故事的理解；也希望在記錄特殊的聲音資料之外，同時提供一種接近他人的聽覺經驗、聽覺記憶的途徑。

Due to its sui generis topography and rich yet turbulent history of development, Taiwan bristles with caves ranging from air-raid shelters to tunnels. Most of them are artificial – for military use or transportation – while few are naturally formed. Their stories become part of our knowledge via the writing system, a process requiring no sensibility involved. Stranded in modernize life, our knowledge has been detached from our perception.

Attempting to experiment on "knowledge" and "perception" as well as reverse their connection and order, this project features the field surveys into these caves around Taiwan and the collection of their stories. Utilizing the natural special sound field, it tries to interpret the first-person auditory experiences in these stories with sound, and represent them through Ambisonic and binaural recording and encoding format, thereby creating an immersive sound journey. With the first-person perspective, the participants immerse themselves spontaneously in this project, and, more important, they narrate the stories according to their own sensory organs. Be-

sides sharing the sound data derived from the special structure of caves, the project expects to create a way to immerse the audience into others' experiences and memories.

REATORS

慈湄 TM LI

為電子音樂創作、劇場音樂工作者、像配樂及DJ。相信聲音能溝通那些「不言說」的事情，使人獲得真正的自由。品多以當代技術駕馭的聲音質性，仍近文化、歷史與社會特色，並著重於破聲音在視覺符號系統中，僅能作為助敘事的功能限制。藉由著重聲音的聽性、空間感與組織性，創造不同的聽經驗與溝通可能。

A DJ, also engages in electronic, theatrical and incidental music. She believes that sound can communicate the indescribable, which genuinely liberates people. Featuring the manipulation of sound itself with contemporary technology, her oeuvre still embodies salient cultural, historical and social characteristics. Apart from using sound as merely an auxiliary of visual/symbolic narratives, she creates various listening experiences and possibilities for communication by highlighting the perception to the property of sound.

▌李慈湄，「凝結歷史殘響：臺灣洞穴的感知敘事」，2020_

TM LI, *Freeze Historical Reverb: the Perceptual Narratives in the Caves in Taiwan*, 2020_

> 「凝結歷史殘響：臺灣洞穴的感知敘事」延續李慈湄自 2019 年於基隆「辶反氵朝」城市劇場行動中，在防空洞演出的創作《未來避難所》，希望更進一步踏查臺灣不同類型的洞穴，一方面也希望利用這樣的特殊空間進行演出。比起在一個黑盒子 / 白盒子空間進行聲音演出，為了要抹去觀眾視覺在演出開始時所進行的「關燈」動作，選擇在本身就沒什麼光線、難以觀看的暗黑洞穴內，更能夠利用原先空間的特質，將除去觀眾視覺的安排做得自然又漸進，且更具有身體感，也因為視覺比重的降低，讓觀眾的其餘感官知覺更敏銳。「洞穴」之於李慈湄，是一個能夠更為貼近其創作理想追求的場域。

> 李慈湄選定新北市平溪區的礦坑與高雄市的鼓山洞，以兩個洞穴各自的歷史背景，作為聲音文本的發展，透過聲音汲取場域的記憶，並轉化為以聆聽為主的感知體驗。在期中進駐發表時，李慈湄先完成一段近十分鐘以平溪礦坑作為主題的聲音創作，並搭配田野採集訪談的文字、影像一起呈現。創作的一開始是山中的蟬鳴聲，隨著腳步踩踏造成的空間回音，讓聽者更加感受到從原本開闊山林逐漸步入洞穴的空間感，接著搭上礦車一路搖搖晃晃地駛進礦坑內部，中段開始則以較為抽象意象的聲響去模擬、轉化礦工身在地底下，身體所承受低氧、高壓下的聆聽狀態。

> 在計畫的尾聲，李慈湄於高雄鼓山洞進行「洞穴聲響創作發表會」，六段演出分別以較為感性的聲響帶出鼓山洞過去的歷史進程，並逐一呈現鼓山洞幾個較具特色的空間：歷史照片回顧、偵查與拷問室、已鐘乳石化直通過去警備總署的天梯、洞穴中央的連通通道，並利用鼓山洞因過去歷史因素，部分洞穴段的牆面則鋪滿隔音用保麗龍薄板，呈現出特殊音場狀態，使聲響的表現有別於一般洞穴中的樣態，時而直襲撲往觀眾，時而迴盪於空間中。演出的結尾與高雄在地的音樂人史旻玠共演，將演出的聲音結合高雄港意象，企圖以聲音將整體意象帶往未來。（文｜馮馨）

觀察報告

馮馨〈聲音場景的取樣、錄製與再建置〉，全文請見
https://mag.clab.org.tw/clabo-article/creators2020-3/

日常截面
Cross-Section of Daily Life

建構在 3D 世界的空間是一個非日常的空間，但有時這些非日常的空間又會讓人聯想到曾經熟悉的空間。究竟虛擬的成像與現實的人與空間之間，存在著什麼連結？是什麼原因使得空間被建立？公共空間與私人空間的區分是因為人的路徑，還是停留時間的長度？黃偉的「日常截面」計畫由此展開，他運用「點雲」（point cloud）的空間掃描技術記錄空間的構成以及人物動態，並進一步運用點雲進行模擬測試，讓空間數據與聲音連動，一方面讓視覺元素所產生的訊號能驅動聲音，另一方面也反過來從聲音的細節、段落變化來驅動影像轉換與畫面細節元素的變化。黃偉所創造的影音從空間出發，又進一步牽連到空間裡的人物，他們在點雲紀錄的空間數據中，甚至變成了遺留在空間中片段破碎的肢體。黃偉將這些影音創作當成日記，記錄自身周遭的環境與空間，用科技的手段重新思考對於事物的感受。

Three-dimensional spaces are non-daily spatial structures which generate certain references to real spaces in daily life. How are the virtual images connected to physical spaces and people? Why have the spaces come to existence? And is the division between public and personal spheres out of the path of the users or the period of time the users use? With these fundamental questions, HUANG Wei's Cross-Section of Daily Life applies a "point cloud" space-scanning technique to record the structures of space and movements of people. It further synchronizes the spatial data and sounds through the point cloud simulations. This approach not only drives the transition of sounds with the signals that are produced by visual elements but also drives the transition of images and adjusts their details with changes to the details and sections of sounds. Starting from the elements of a space, the images and sounds that Huang created also influence the people and objects within that space, rendering them shattered extremities that are left in the space in the point cloud spatial data. Huang views his audio-visual creations as diaries that document the environments and spaces around him and reflect his perceptions of those surroundings through technological means.

REATORS

黃偉 HUANG Wei

畢業於國立臺北藝術大學新媒體藝術學系，擅長音像演出、互動影像設計、空間裝置。他的創作圍繞在對生活的觀察，並透過對於物件的直觀想像，將非日常的畫面帶入日常中。其關注焦點為如何藉由影像與聲音強調現實的身體感與既視感，並致力於電腦演算的影像與聲響結合進行演出。

Graduated from the Department of New Media Art of Taipei National University of the Arts, HUANG Wei specializes in audio-visual performances, interactive image design, and installations. Based on his observations and intuitive perceptions of life, Huang's work integrates uncanny images into the scenes of daily life, bringing out a sort of *déjà vu* experience in seeing and sensing through audio and visual manipulations generated by algorithm-based performances.

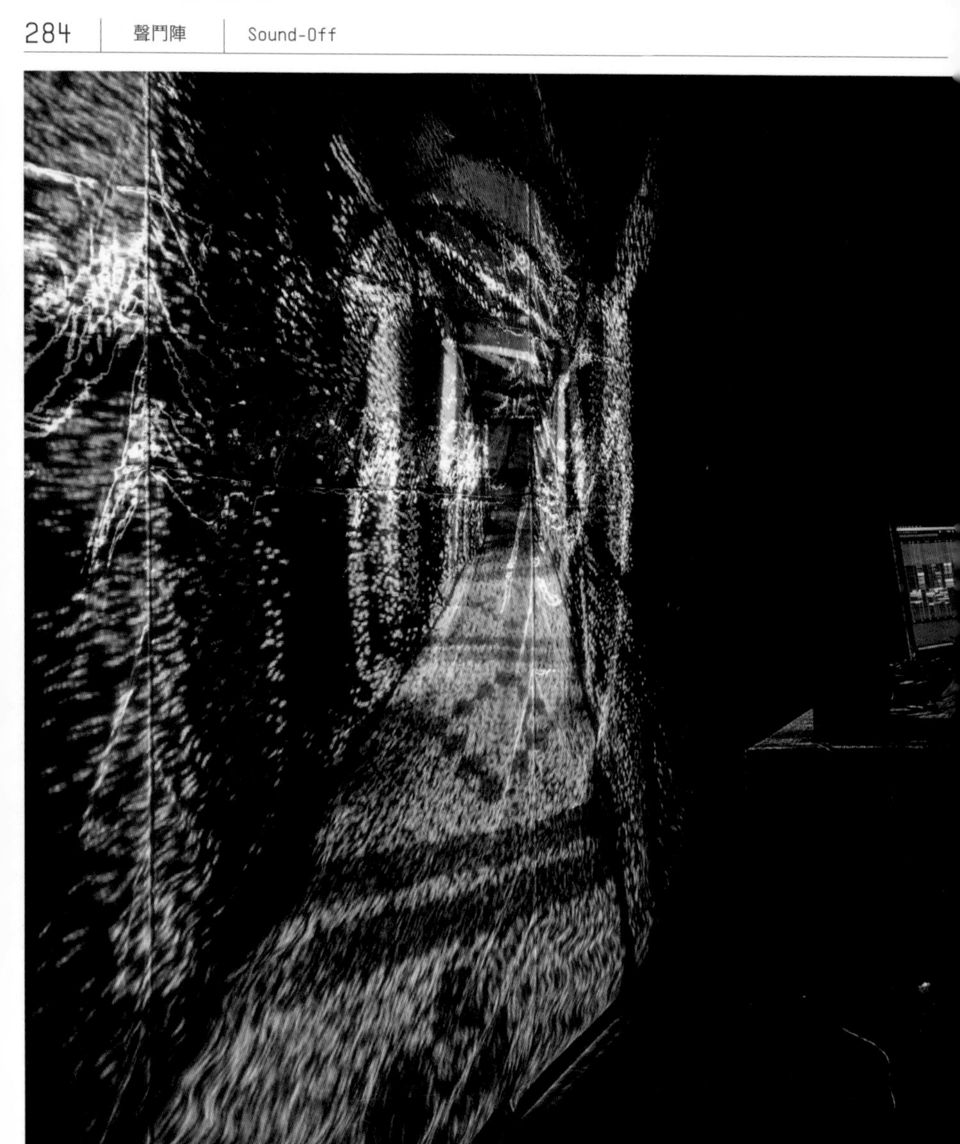

▌黃偉，「日常截面」，2020_

HUANG Wei, *Cross-Section of Daily Life*, 2020_

人工自然——臺灣聲響實驗室沉浸式體驗研究與創作計畫

Artificial Nature – Research and Creation of Ambisonics System in Taiwan Sound Lab

這項計畫結合 Ambisonics 環繞聲響技術與器樂演奏，探索空間聲響、感官期望、作品與觀眾之主客關係，從音樂聲響創作延伸出各項創作議題，包括觀眾主觀知覺、真實聲響與人工殘響之交織對比、聲響空間性的自然常規與反常規、空間聲響作品與視覺訊息的關係等。作為此計畫的階段性呈現，黃苓瑄完成了一個多聲道聲音錄像裝置，旨在錯置觀者的視覺與聽覺，讓我們在面對自然物理視覺現象時，遭遇違反聽覺習慣的聲音；於臺灣聲響實驗室的現場演出，則是在一個 Ambisonics 音場中組合表演者的身體動態與聲響的空間物理。這兩種形式都以「音樂與重力」為主軸，圍繞著視覺和聽覺訊息之對抗與互動。

This project incorporats the full-sphere surround sound technology of Ambisonics and instrumental performance to explore the relationships between space sounds, sensual expectations, sound works, and the audience. HUNAG Ling-Hsuan has further raised issues relating to music and sound creations, including comparisons of subjective perceptions of audience, real sounds, and artificial reverberation, the natural and unnatural reflections of sound spatiality, and the relationship between spatial sound work and visual messages. As an embodiment of the project in its current phase, she created a multi-channel audio-visual installation that aims to dislocate the audience's vision and hearing by presenting sounds against our hearing habits when encountering natural physical phenomena. Meanwhile, the live performance in the Taiwan Sound Lab was an attempt to collage the performer's body movements with the spatial physics of sound. Both of these presentations focus on music and gravity to capture the confrontation and interaction between visual and audio messages.

REATORS

琴瑄 HUANG Ling-Hsuan

曲家。其音樂探討聲音與內外感官、律與意外的多面維度思考。近來於品中亦嘗試音樂與儀式元素結合，自然音響的人工式詮釋，作品包含奏、室內樂、管弦樂、電子音樂、裝、音樂劇場等形式。自國立臺灣師範學音樂系畢業後赴歐，於德國柏林術大學取得作曲碩士文憑與卡爾斯厄音樂院作曲最高文憑，此後負笈蘭海牙皇家音樂院專攻電子音樂學。曾於2018年獲德國 Deutscher usikwettbewerb 作曲首獎。

HUANG Ling-Hsuan is a composer exploring the relationships between music and perception, as well as ritual elements and daily life. She earned her bachelor's degree in Composition from National Taiwan Normal University, master's degree from Berlin University of the Arts, and a Konzertexamen diploma in Composition from University of Music Karlsruhe, where she studied under Wolfgang Rihm. She has also pursued her interest in Sonology at The Royal Conservatory in The Hague. Huang creates electronic music and music for installations, in addition to instrumental compositions for solo artists, chamber ensembles, theater companies, and orchestras, drawing inspiration from the senses and musical texture of regularity. She won the German Music Competition (Deutscher Musikwettbewerb) in 2018.

▌黃苓瑄，「人工自然——臺灣聲響實驗室沉浸式體驗研究與創作計畫」，2020_

HUANG Ling-Hsuan, *Artificial Nature - Research and Creation of Ambisonics System in Taiwan Sound Lab and Audio-Visual Presentation with Head-Mounted Display*, 2020_

Signals：實驗通信 CREATORS 2020-21

展期 Date ▶

2021/2/19

|

2021/3/14

2 安魂工作隊 Libera Work-Gang，2019-2020_
《的確是存在於二十世紀》，錄像、版畫、混合媒材裝置
Did Exist in 20th Century, video, woodcut, mix-media installation

3 展覽開幕現場 The exhibition's opening scene

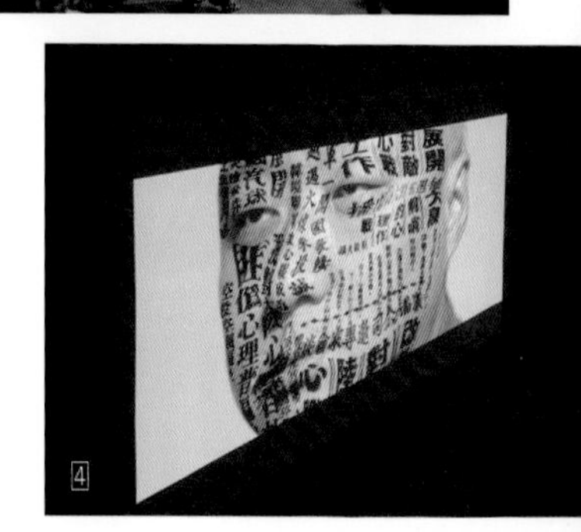

1 引爆火山工程 Engineering of Volcano Detonating，2018_
《負地理學》，錄像、文件
Neg-geography, video, documents

許哲瑜 HSU Che-Yu／陳琬尹 CHEN Wan-Yin，2020-2021_
《綠頭鴨之死》，錄像裝置
The Unusual Death of a Mallard
video installation

4 遠房親戚實驗室 Lab of the Distant Relatives，2020_
《芥面：正體中文版 1.0》，網站、錄像、文件、植物
Intergrass: Traditional Chinese Edition 1.0,
Websit, vedio, documents, plant

5 展覽現場 Exhibition scene

1 走路草農/藝團 Walking Grass Agriculture，2020_
《噥噥時間交換所》，錄像、手稿、壁畫、混合媒材裝置、公眾活動
Murmur Time Project, video, drawings, wall painting, readymades, mix-media installation, puclic event

2 黃鼎云 HUANG Ding-Yun，2020-2021_
《空耳乩》（神的樓所 GIR 階段紀錄），雙頻道錄像裝置
*Misheard Medium (an acting documentation of -
in Residency)*, double-channel video installatio

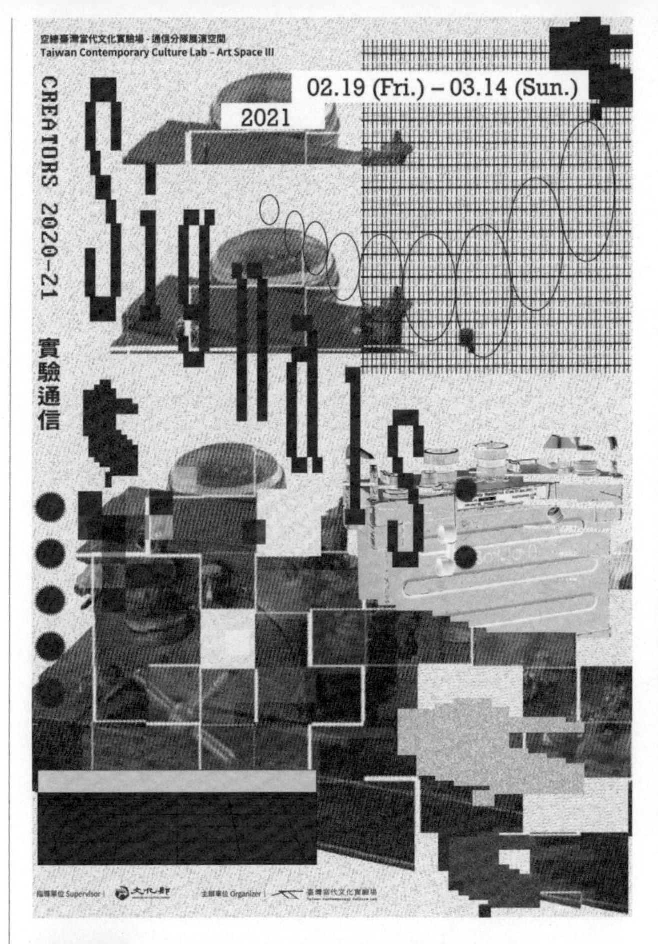

空總臺灣當代文化實驗場 - 通信分隊展演空間
Taiwan Contemporary Culture Lab – Art Space III
CREATORS 2020-21
實驗通信
Signals
2021
02.19 (Fri.) – 03.14 (Sun.)
指導單位 Supervisor | 文化部
主辦單位 Organizer | 臺灣當代文化實驗場

藝術家
ARTISTS

引爆火山工程、她的實驗室空間集、安魂工作隊、江之翠劇場、吳秉聖、李慈湄、走路草農 / 藝團、施懿珊、許哲瑜 / 陳琬尹、陳志建、黃苓瑄、黃偉、黃博志、黃鼎云、遠房親戚實驗室、羅懿君

Engineering of Volcano Detonating, Her Lab Space, Libera Work-Gang, Gang-a Tsui Theater, Wu Ping-Sheng, TM LI, Walking Grass Agriculture, SHIH Yi-Shan, HSU Che-Yu/ CHEN Wan-Yin, CHEN Chih-Chien, HUANG Ling-Hsuan, HUANG Wei, HUANG Po-Chih, HUANG Ding-Yun, Lab of the Distant Relatives, LO Yi-Chun

文化實驗之關鍵不在展示「成果」，而在揭露潛在的「尺度」

——從「CREATORS 計畫」談創研實驗的可視化問題

文｜王聖閎．圖｜臺灣當代文化實驗場

三種時間的尺度

以下三個獨立事件，分別涉及人類活動中至少三種截然不同的時間進程，同時也與人們對事物會發生變革所需的時間、速度及頻率的基本認知有關。本文將以此為起點，開啟關於「文化實驗」的一系列思考：

HFT (High Frequency Trading) 又名高頻交易，是指充分運用自動交易的設計，從超出人類反應時間的市場變化中（譬如極微小的賣出和買入價差）設法獲利的一種交易方式。2010 年 5 月 6 日發生的著名「閃崩」事件（Flash Crash），即是源於某個交易員試圖透過高頻交易操控股市，使得當天美國道瓊工業指數在短短 36 分鐘之內下挫了 9%，令市場損失共約 8,620 億美元，引發軒然大波。

日治時期，臺灣曾經是主要的檜木輸出地。阿里山曾設有不少伐木廠，將重要的檜木資源運至日本蓋神社和廟宇。二戰後，臺灣的林業政策仍繼續伐木，致使大量的原始森林遭到破壞。檜木的生長速度緩慢，平均約 350 年才能長高 1 公尺。因此許多自日治時期之後就從臺灣土地上消失的山林資產，直到今天都無法復原。

2019 年，NASA 科學家詹姆士．歐杜諾修（James O'DONOGHUE）因為學習動畫製作之故，在網路上分享了自製影片，以清楚易懂的方式解說光速在宇宙中傳遞需要耗費多少時間。當影片以實時（real-time）速度呈現光從地球抵達火星，最快仍需要至少 3 分 2 秒之久時，一般大眾不只獲得一個理解光速究竟多快（或多慢）的簡明視覺化形式，同時更獲得了具體感受宇宙空間究竟多廣闊的有效途徑。

上述三個獨立事件分別在不同的時間尺度下進行：毫秒、百年、分鐘。股市崩壞只需要一眨眼的功夫、樹木生長需要數百年甚至千年以上的等待，而以光速進行的星際通訊，遠比人們想像中的更為漫長。儘管這三者耗費的歲月天差地別，但只要從適當的尺度理解，我們便能對其變化進程有合理的估算和期待。相反地，若是缺乏這種必要的速度學認知，將宏觀世界之變化壓縮在人類文化活動的有限視野下理解，我們就可能將本屬不

[1] 關於這種從尺度規模之視角，反思人文學科對於環境、生物、文化及歷史的理解限制，以及強調不同尺度之間無法簡單化約的觀點，可參閱：林開世，〈期待一個不同尺度的未來：給 2020 年台大人類學系畢業生〉，《芭樂人類學》網站，2021.01.28 瀏覽。

同尺度規模的事物，混為一談。甚至，忽略了即使在人類短暫的歷史中，每一種物質文化（及其所對應的生產技術、觀念）所需的時間進程都不盡相同，以至於我們總是制訂出過度追求短效、倉促的未來計畫。[1]

舉例來說，當我們在藝術領域中談及「文化實驗」的概念時，究竟是將之放在何種時間尺度下理解的？若更具體就創作中的實驗來談，摸索一項全新的創作材料需要多久時間？將某種陌生的展呈技術開發成熟並普及化，又需要多少時間？一個藝術家自認必須潛沉多久，才能醞釀出她 / 他下一個展演計畫的雛形？而其認知相較於國家藝文補助單位對創作週期的想像，又存在多大的落差？如果每一種技術、材料、形式和觀念的革新，所需的社會條件各不相同，過於扁平狹隘的展演製作期程會不會將我們侷限在一個「當代性」的迷障裡？或反之，當我們強調所有創作計畫都仍在持續演進，並且在做展覽時大量依賴文件、調研紀錄、草圖或測試版本，會不會因此反而過度濫用了「work in progress」的意義？

總而言之，回到最根本的問題：對「文化實驗」的深耕培育而言，我們是否思考過哪一種時間尺度（與速度）概念才是相對合理的？

反對速度利維坦主義

特別是身處在 COVID-19 這場世紀疫情之下，上述的難題格外值得深思。因為病毒極大程度改寫了我們對於「速度」的既定認知。不少國家因為一開始的掉以輕心，如今更加集中掌控所有與生命存續（和監管）息息相關的技術介入力量。同時，也變得更為重視生產力 / 研發力 / 組織力的即時配置，以及對局勢變化的快速因應能力，以至於，「速度」兩字被大幅抬升到幾乎是政治正確的地步。它既是核心的文化價值，也是總體國力的展現。某方面來說，COVID-19 像是促成了一台以全球為規模的超大型加速器，使所有國家皆籠罩在一股「速度利維坦主義」（Speed-Leviathanism）的氛圍之中。它一方面掀起這波同時席捲經濟、文化、社會與政治層面的巨大風暴。另一方面，也似乎在人們心裡暗地置入一種觀念：意即假定這種例外狀態下的生活，一切皆可加速。而當例外狀態逐步篡位，最終成為我們的生活常規時，那些無法加速的事物將難以適應未來挑戰，注定遭到淘汰的命運。

[2] Paul VIRILIO, *Speed & Politics*, New York: Semiotext(e), 1986, p.69.

簡言之，速度利維坦主義無法忍受資源的無端耗費，更不可能接受人力和時間的虛擲，最終卻無法換取任何成果。即使是所謂的「創新」或「研發」，也必定是工具理性和目的性導向的。所有一切都能以國家總體競爭力之名，在技術官僚的算計邏輯下落實為可視化的成果──那些雖然短效但具體可行的成果。據此，「防疫視同作戰」遠遠不只是口號，而是將戰爭修辭確實轉化為政府機構的行動信條，並且強調施政成功與否的訣竅，恰恰就是速度和效率。相反地，若是無法進一步提升作戰速度，盡快挺過疫情並恢復經濟活力，那些反應過於遲鈍的國家政體勢必會在這場生命政治的競速賽局中敗下陣來。就此而言，COVID-19 帶來的不僅僅是資本主義加速機器再一次突變的絕佳誘因，更是一場速度利維坦主義思想正式上位的全球化佈告：它變態地要求所有國家都應該成為保羅．維希留（Paul VIRILIO）口中的「競速政體」（dromocracy），而非民主政體；國家成敗如今只與競速學（dromology）相關，而非傳統的政治學策略。[2]

但值得我們反思的正是這點：「文化實驗」的進程，難道也必須受制於速度利維坦主義式的時間邏輯，綁縛在偏狹的社會加速（social acceleration）思維之下？事實上，對「實驗」兩字提出迥然不同的進程想像和發展步調，正是臺灣當代文化實驗場（C-LAB）最初極力向外界證明的一點。作為新型態的文化機構，它自成立以來面臨的最大壓力，即是國家主政者和民意代表希望立即看見「具體施政成效」的考核思維，以及空總周邊開發主義勢力的不斷進逼。平心而論，為了順應這些外部要求（或者，為了替內部實驗性計畫爭取相對自由的空間），近期的 C-LAB 也大量策辦與一般藝術場館無異的展演活動，搭配幾乎可說是過量的工作坊、講座、論壇、分享會，努力創造聲量和關注度，使一般民眾注意到這個地方持續有活動在產出。

但真正能夠標誌 C-LAB 特質，或最能彰顯它作為「文化實驗場」的一點，恐怕要屬「CREATORS 創作／研發支持計畫」（簡稱「CREATORS 計畫」）的獨特設計。因為該計畫與一般駐村或創作補助機制最明顯的差異，就是它全然不是以成果或績效考核為導向，也不會要求在進駐計畫結束之時，藝術家和其團隊必須拿出某個明晰可辨的「創作實驗成果」。除了提供常規的資源、空間、軟硬體設施，「CREATORS 計畫」更為重視專家顧問之媒合，以及評論觀察員等陪伴機制的支持，著眼於搭建一個創作思想得以站穩腳步的立基點。也就是說，「CREATORS 計畫」的初始設計，同時解除了兩個纏繞著臺灣當代藝術的緊箍咒：一是假定創作的實驗性探索，必定以展覽或作品製作為其終點。二是假定創作實踐必定是「可展示的」，或必然朝向可視化的成果努力。它表面上雖然仍以「計畫」之名，但在藝文補助環境日益科層化（所有人都沉淪於無止境的申請表格與階段報告），同時也高度計量化（所有獲補助／未補助紀錄必須一目了然）的今天，「CREATORS 計畫」多少保留了一點藝術系統內部的不穩定性——它無法輕易化約為僵固的時程進度表，難以削足適履地卡入一般性的計畫性思維裡；它甚至允許創作實驗的某些區塊，既不可展示，亦無法被加速。要言之，「CREATORS 計畫」的推動，某個程度上代表我們依然願意相信「文化實驗」本身就是極度耗時、耗力的。它不僅短時間之內難以立竿見影，更存在「實驗會失敗」的可能性，從而對這個瘋狂競逐展演活動輪替速度的當代藝術世界，帶來一定程度的擾動。

特別指出這點，並不是為了說明「文化實驗」僅適用於純粹耗費式的藝術生產邏輯。畢竟我們早已遠離 1990 年代「台北國際後工業藝術祭」那種生猛有力、可不計一切代價的生產場景。正好相反，「CREATORS 計畫」的存在恰恰能幫助我們反思：這個看似勢不可檔的「大企劃書年代」還存在哪些缺陷和

[3] 朱貽安，〈綜看空總：專訪C-LAB空總臺灣當代文化實驗場執行長賴香伶〉，《ARTouch》網站，2021.01.28瀏覽。

[4] 摘自「實驗通信 CREATORS 2020-21」展覽手冊文字。

盲點？我們會不會其實根本不懂怎麼做真正長遠的規劃？不知如何給予實驗、試煉、探索、思考真正充裕的時間，讓創造性的事物得以自然醞釀和成形？

計畫的不透明性，與知識的可視化迷思

當然，目前「CREATORS 計畫」本身並非毫無挑戰。它一方面有資源分配上的壓力，另一方面，由於收錄的創作實驗類型既多又廣，舉凡生物藝術、新媒體藝術、聲音藝術、科學調研、歷史踏查，甚至地方創生，皆在該計畫支持的守備範圍之內。但嚴格而論，除非 C-LAB 本身的內部編制人力更為多元，否則並不容易細緻照顧到如此多面向的跨領域內容。也就是說，如果「CREATORS 計畫」僅僅扮演資源平台與初階媒合者的角色（儘管妥善做到這點已實屬不易），它就只能是短暫且缺乏延續性的進駐機制。縱使雨露均霑卻難以長久積累，形成真正具有明晰脈絡的「實驗文化」；它可以激起討論火花，但因為缺乏縱深視野以及對個別計畫的長期性追蹤，難以就地形成為某種文化聚落（settlement）。在此，「settle」之意本就是使人得以安頓，使創作生活的節奏得以穩定發展，並且使「嘗錯」和「試煉」的環境能夠落實。這些條件都不是頻繁抽換、輪動的運作模式可以提供的。

除此之外，「CREATORS 計畫」真正面臨的根本問題，在於它始終難以擺脫「可視化邏輯」的幽靈，一再地捲進「於非典型空間裡做展覽」的侷限，與其初始設計有所抵觸。[3] 這份不得不然，充分突顯在近期推出的「實驗通信 CREATORS 2020-21」裡，並在策展人游崴這段文字[4] 中表露無遺：

> —— 展覽作為一項發表機制，捲入了整個文化體制對於透明性的渴望——展示讓計畫變得可以評估、可以測量。文化實踐在規章中被錨定、在系統的凝視下變得透明，成為國家總體競爭力的一個計算項目。在另一端，獨立的文化生產者則進入了寫計畫、申請、執行、記錄、階段報告、結案報告與等待尾款的工作週期。然而，每個計畫自身的生命週期卻不僅於此，背後支撐著的，是對於特定主題一種時間跨距更大的熱情，而真正恢宏的計畫總是難以結案，或無法結案。

顯然，策展人也相當清楚，「文化實驗」必須展示這點，更多是出於國家藝文科層體系本身的可視化要求。因此這些調研速度不一、執行時間跨距也迥異的進駐團隊，被迫必須在齊頭式的截止期限內，溫馴地端出各種檔案文件、草稿素材，或者裝置原型充作可觸、可感的「成果」（美其名，一個階段性的實驗「節點」）。在支援系統即將關閉之際，16 組進駐團隊幾乎都難以大膽地為自身辯護，彰顯其計畫「既不可展示，亦無法被加速」的一面——也就是藝術實踐有強度但純粹耗費的一面，或者「文化實驗」本身不透明的一面。以至於，「實驗通信」展雖然暫時滿足了外界的好奇心，但由於想要快速窺探創作實驗過程的凝視，本身就是視野扁平且趨於即效性的，越是屈從於這種可視化邏輯，展覽就越是會落入碎片化、分段化呈現的窘境，難以顧全每一個進駐計畫各自的背景資訊脈絡。儘管每一組團隊所耕耘的議題都是有趣的，但由於展場空間相對侷促，計畫之間並不容易調和彼此的殊異屬性，展呈上又相互干擾，觀眾很難透過浮光掠影的檔案碎片，充分掌握作品或計畫本身的真正潛能（potentiality）。

上述問題的癥結其來有自，並非一個 C-LAB 就能單獨解決。因為可視化邏輯的根源，至少可以追溯到「2012 台北雙年展：現代怪獸／想像的死而復生」。在該屆雙年展之後，臺灣當代藝術不僅啟動對歷史檔案的熱切凝視，更能清楚觀察到一波視覺藝術展覽形式的「博物館／博物學化」傾向，為日後「田野」兩字的熱議鋪路。儘管，這波持續至今的創作轉向有其深遠意義，但「副作用」也相當明確：展覽如今被預設是一種快速的知識交換介面和平台。無論策展人和藝術家展出的是已完成的作品，還是能充分體現其計畫之過程性的田調資料、地圖、手稿筆記，展覽往往以「知識的即時交換」作為其隱含前提。越是博物館化的展呈形式（它們多半毫不節制地塞滿大量的檔案文件），就越是假定觀眾在展場閱覽一圈之後，就能理解並帶走這些藝術知識。此即可視化邏輯的基礎原型。

然而，這種假定一切皆能轉換成可見／可感／可交換形式的預設，往往忽略了在田野踏查，或者實驗探索的「第一現場」裡，存在著許多不可見的「身體性知識」。它們或許可以透過文字說明讓觀眾初步意會，卻無法被立即給予；它們需要相當時間的沉澱、醞釀和體會，甚至必須「活入」（live in）才有辦

法習得。當代藝術如果傲慢地假定，這些僅能在不同時間進程和步調中獲得的緩慢知識，全都能順理成章地轉置到展覽空間的「第二現場」（那誇誇其言的「讓不可見變得可見」），甚至以為第二現場就足以取代第一現場，我們恐怕將陷入一種「再現性思維」的展示暴力而渾然不知。[5]

要言之，倘若無法對這種「一切皆可置入展場」的隱性預設有所警覺，我們自然會濫用任何能彰顯「work in progress」的表達形式，並且難以抗拒可視化邏輯的誘惑。而這種總是趨於速成、短效的傾向，不僅難以深化當代藝術的知識生產體系，同時也會讓創作者對其實驗計畫的時間尺度想

[5] 關於「再現性思維」的批評，請參閱拙文：〈田野調查，一個當代藝術實踐方法論的初步探問〉，原刊於《典藏．今藝術》，253期，2013.10，頁114-117。

像，徹底壓制在如影隨形的展示暴力之下。據此，「文化實驗」非但無法成為我們反省速度利維坦主義的最後陣地，深陷系統透明性迷思的我們，連同「文化實驗」所能開啟的新一輪創作倫理暨方法學探討，反而都在日漸被人們視為理所當然的社會加速氛圍中，成了早夭的生命。

設想「文化實驗」的內蘊化形式

嚴格而論，這裡的癥結並不是我們只能在「展示／不可展示」之間二擇一。真正的問題在於思考：「文化實驗」真正適合對外展露的究竟是什麼？譬如此次在「實驗通信」中展出的「引爆火山工程」，現場出示的寥寥幾筆文件和活動照，根本不足以說明這個年輕團隊來回「橫越」（transverse）地質學、考古學、天文學，乃至於飛碟研究領域，找尋科學／非科學知識之間的縫隙與辯證關係的企圖心。該計畫的潛在價值遠遠不是那些能置於展場內的影像或物件，而是如他們的系列講座所呈現的，一種既有學科架構不易促成的知識社群挪移。或者，如吳秉聖的「投聲計畫」和李慈湄的「凝結歷史殘響：臺灣洞穴的感知敘事」計畫。前者僅能展示相關草稿，以及其反覆測試的喇叭陣列裝置原型。觀眾可以感受瞭解裝置和技術的「外貌」，卻未必能在 C-LAB 有限的空間條件下知其所以然。類似地，後者透過文史物件盡心布置的聆聽室，恐怕也只能給予一個概略的「類現場情境」，卻不足以提供充分資訊，使觀眾進一步瞭解 Ambisonics 與洞穴錄音背後的技術思想，遑論交代技術本身與臺灣礦業或軍事遺址之間的關聯。也就是說，吳秉聖與李慈湄各自的計畫當然都有做成展覽的潛力，但很可能不是現在；現階段的裝置或體驗式呈現，或許滿足了文化體制對實驗過程的透明性渴望，但卻完完全全受制於一種頗為扁平的日程投射認知，汲汲於找出觀眾可立即帶走的知識形式。[編按]

必須特別強調，上述問題並不能簡單歸咎於展呈設想上的失誤。主因是這些計畫某個程度上都受限於可視化邏輯，因此它們往往不是在過早的探索階段就被要求進入展示思維，就是被迫呈現其「第一現場」，公開「文化實驗」本身不透明的暗面。換言之，這裡的癥結並不在於「文化實驗」能否藉由主張自身的不可見性，以規避公眾檢驗與監督的問題。[6] 而是我們很可能誤解了「文化實驗」的核心工作其實不是創作思想的實際展

[編按] 兩位藝術家於「實驗通信」展場中的靜態展示之外，展期間亦有相應的演出活動，包括吳秉聖的聲音裝置體驗場次《投聲系列：校音一》與李慈湄的聲音影像演出《回聲／反射 I》。

[6] 也就是說，這裡的問題比較不是吳牧青所言的藝文界「不愛被管」的慣習。此種慣習確實存在，但如今的問題是我們並不真的明白，如何在文化體制要求的透明性，與創作實驗推進的既有節奏之間取得平衡。見：吳牧青，〈彌留的空總素地，藝文界的七寶奇謀〉，原刊於《典藏・今藝術》，337 期，2020.10，頁 102-103。

現（manifestation / demonstration），而是創作之力的儲存、蓄積與蘊藏。在此，「實驗」兩字的真正要旨，是為創作思想找到一種可行的內蘊化形式（implication），使其力量得以恰當地收束和沉澱，並於未來適當的時機點綻現。而以此為核心價值的機構，從來都不應該在一檔接一檔的展演製作中不斷地消耗自身氣力。

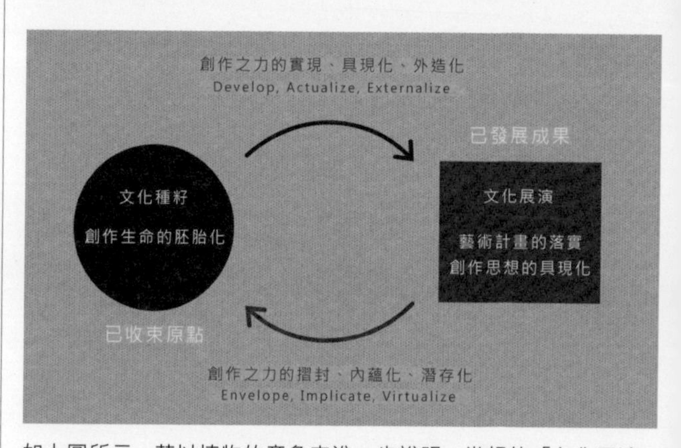

如上圖所示，若以植物的意象來進一步說明，常規的「文化展演」就像是一株消耗土壤沃力和養分而長成的葡萄藤蔓。周遭的陽光空氣水等環境條件，決定了它是否可以長得完善漂亮。「文化展演」的要旨，是透過吸收先前已積累的文化力量，使創作者們的特異想像得以實現，創作思想得以具現化（actualization）。

其基本思路是透過層層規劃，使事先擬定的藝術內容得以逐一「展延」（developing）開來，落實成可觸可感的最終成果。但相較之下，「文化實驗」則猶如一顆儲存能量與基因序列的葡萄籽，其動力是思考如何將創作之力的「多」（the multiple）蘊藏於「一」（the one），是為創作生命找到胚胎化的原點，一個屬於「摺封」（enveloping）的動勢。「文化實驗」的關鍵不在於長成姿態各異的葡萄藤蔓，而是找出最能保有藝術實踐之純粹潛能與潛在性（virtuality）的適當形式。據此，倘若「文化實驗」必須對外顯影，至多也只是呈現某種可由「一」窺見「多」的內蘊化形式——它們可以是眾多有強度，但無確切成果展出的非正式交流，也可以是有思想激盪，但無限定期程的關係生產——重點是與「一切皆可置入展場」的具現化邏輯拉開距離，並逐步形成一種以內蘊化為核心工程的常態性聚落。或更精簡地說，試著推動一種文化風氣（ethos）。正如「CREATORS 計畫」的原初精神，時時允許與現實保持某種「間距」（但亦步亦趨），不急於實現、不倉促展示，始終將自身維持在潛存狀態之中，並逐步拼湊出一塊創作者與研究者可以慢慢做、慢慢想的感性「精神角落」。

在各大藝文場館瘋狂製作高密度的展演活動，各藝術大學也陷溺在「卓越」、「願景」、「創新」等科層語言迷思的今天，試問還有什麼地方能夠逆風抗拒速度利維坦主義的席捲，並專注於文化力量得以蓄積的「潛存化」（virtualization）工程？這原本是 C-LAB 所能彰顯的一大特點。但如今也深深捲入生產速度學風暴的它，恐怕難以再對文件展形式的氾濫有所反思；它也塌陷成另一個與其他場館無異的當代藝術展演中心。

平心而論，只要 C-LAB 展覽做得越多，它就離「CREATORS 計畫」的原初精神越遠。因為任何展演製作都必定是人力和心力的大量耗費。即使今日 C-LAB 的工作團隊編制充裕，只要它持續以具現化的思維來面對「文化實驗」，它的第一線人員就必須面對過量的發表壓力，並且在事倍功半的空間條件下，做到不輸常規藝文機構檔期數量的展演活動。但於此同時，他們仍必須在極其壓縮的時程內，兼顧各項實驗計畫的主持、招募和看顧。這種蠟燭兩頭燒的工作模式，對任何成熟的團隊而言，都是一種力量的嚴重消磨，而非深耕和奠基。同時它的被迫「推陳出新」卻可能間接給予外界錯誤的認知跟期待，以不合適的時間尺度與發表期程來理解、體驗「文化實驗」的淺層內容。如此分身乏術的 C-LAB 無疑是工作過量的。

這是以為藝術實踐勢必只能走在「實線」上（意即，創作不可以走在不實現的「虛線」上）的臺灣當代藝術，始終解除不了的一道緊箍咒。然而，當它深信一切都必須能被藝文科層系統所登載、管理、追蹤，以及核銷時，它在具現化與潛存化之間的進退失據，卻首先由 C-LAB 來承擔。這點，恐怕才是 C-LAB 如今面臨的最大考驗。

作者簡介

王聖閎

國立臺北藝術大學美術學系博士，現為國立中央大學藝術學研究所專任助理教授。主要研究興趣包括當代錄像、行為表演及身體性創作；關注當代加速度政治下的生命政治課題，聚焦科層機器對身體行動和耗費模式之影響，並且在此基礎之上，持續反思在地藝 術理論的生產與擴延。

● 參考資料

1 引爆火山工程網站：https://spacestudio1219.wordpress.com/。

2 沈克諭，〈不在此時：時差的潛勢〉，《CLABO 實驗波》，2021.03.08。

3 馮馨，〈聲音場景的取樣、錄製與再建置〉，《CLABO 實驗波》，2021.03.02。

4 王聖閎，〈科層時代的藝術生產之境——從「我們是否工作過量」的策展問題意識談起（上）〉，《CLABO 實驗波》，2019.10.13。

5 王聖閎，〈科層時代的藝術生產之境——從「我們是否工作過量」的策展問題意識談起（下）〉，《CLABO 實驗波》，2019.10.13。

The key to cultural experimentation is not to exhibit "results," but to expose the latent "scale"

——On the visualization of creative research experiments in the CREATORS program

TEXT _ WANG Sheng-Hung, (Translated by Jack WANG)
PHOTO _ Taiwan Contemporary Culture Lab

The three scales of time

The following three independent events take place over at least three disparate progressions of time in human activity. At the same time, they relate to the basic human conceptions of time, speed, and frequency. This article starts with an examination of these events to embark on a series of thinking on the "cultural experiment":

High Frequency Trading, or HFT, designed to fully leverage automation processes, attempts to profit from fluctuations of prices too fleeting for human reflexes to act on (through miniscule margins of buying and selling) in the stock market. The notorious flash crash of May 6, 2010 was the result of a single trader trying to manipulate the stock market through HFT, who in the course of 36 minutes caused a 9% fall in the Dow Jones Industrial Average Index, incurring a loss of 862 billion US dollars in share values and social unrest.

During the Japanese colonial era, Taiwan was one of the main exporters of the hinoki cypress, processed at numerous lumber mills built throughout Alishan, sending the all-important resource off to Japan for construction of Shinto shrines and Buddhist temples. Post-WWII, Taiwan's lumber industry continued to operate, causing rapid loss of old-growth forests. The hinoki cypress grows at a glacial pace, at just 1 meter per 350 years. Thus, the natural heritage lost in Taiwan since the Japanese colonial era has not been replenished to this day.

In 2019, NASA scientist James O'DONOGHUE was learning how to create animated pictures, when he shared his creation on the internet, explaining in relatable terms how long it takes for light to travel in space. When it was demonstrated in the real-time video that it takes 3 minutes and 2 seconds for light to reach Mars from Earth, the public not only gained a visual means of understanding of how fast (or slow) light speed is, but also found an effective way to conceptualize how vast space can be.

These three independent events occur under different time scales: in milliseconds, over centuries, and in minutes. A stock market crash happens in the blink of an eye, trees grow over hundreds of years or even thousands, and interstellar communications are much less instantaneous than we imagine it would be, even at lightspeed. Despite vastly different scales of these timelines, we can get a reasonable estimate and expectation of these progressions if we observe each one at its appropriate time scale. Conversely, if we lack an appropriate sense of chronology, considering contexts of a macro scale within the limited perspective of human existence and cultural activities, we will likely confuse different events which can only take place on their respective time scales. Furthermore, we may overlook the fact that within the short span of human history, each material culture (and corresponding production technologies or concepts) takes place through different progressions of time, so that we often make plans that are short-sighted and rushed.[1]

[1] Regarding a scale-specific perspective towards the reflection of the limitations of humanities' understanding on the environment, biology, culture, history, and the irreducibility of scale, refer to: LIN Kai-shyh, *Looking forward to a future of different scales: to the graduates of the NTU Department of Anthropology*, Guavanthropology website, accessed January 28, 2021.

For example, what time scale should we assume when considering the concept of "cultural experimentation" in art? Specifically, during the experimentation phase of creative processes, how much time does it take to fully grasp new material for creative purposes? How much time does it take to polish and popularize a new technique of exhibition and representation? How long does the artist think she or he needs in gestation before an exhibition plan can be hatched? And how does the artist's conception of the creative cycle differ from that of the sponsoring national arts and culture offices? If the evolution of each technology, material, format, and concept relies upon different social conditions, then would a singular and constrained exhibition production cycle trap us within the confines of "contemporancity"? Or conversely, if we claim that all creative projects are continuously evolving and rely on copious amounts of documentation, research notes, sketches, or mock-ups in an exhibition, would we not be guilty of abusing the term "work in progress"?

In summary, the fundamental question remains: in the cultivation of "cultural experiment," have we given thought to what a reasonable concept of time scale (or velocity) may be?

Against Speed-Leviathanism

These questions bear a deeper reflection, especially in times of the historic COVID-19 pandemic. This is because the virus has largely rewritten our preconceptions of "velocity." Many countries had initially underestimated the situation, later coming around to tighten the grip via all manner of technological interventions concerning the preservation of life (and its regulation). At the same time, countries placed greater emphasis on the timely allocation of production/research/governance, and the ability to quickly respond to changes in a situation, so that "speed" had come to signify politically correctness. It has become both the core cultural value and an indicator of national power. In some ways, COVID-19 has produced a world-scale super-accelerator, enveloping all nations in the wake of speed-Leviathanism. On the one hand, it stirs up a combined economic, cultural, social, and political superstorm. On the other, it plants a seed in the hearts of citizens, with the assumption that within a life of exceptional circumstances, <u>everything can be accelerated.</u> And as the extraordinary circumstances take hold, becoming established as a fact of life, the things which cannot be accelerated struggle to adapt to future challenges, and are destined for obsolescence.

Simply put, speed-Leviathanism cannot allow for a futile expenditure of resources, especially not a lavish waste of manpower and manhours without recompense. Even the so-called "innovative" or "developmental" activities must be instrumentally rational and purpose-driven. In the calculated logic of the technocracy, everything becomes visualizable results in the name of national competitiveness — achieving results that are short-acting but practicable. With this, the slogan "a war on the pandemic" is more than a phrase, but an actual appropriation of military rhetoric and transforming it

[2] Paul VIRILIO, *Speed & Politics*, New York: Semiotext(e), 1986, p.69.

into a mandate for governing bodies, whose legitimacy, coincidentally, is measured in speed and efficacy. Conversely, a failure to act quickly to maintain and restore economic performance will surely be a mark of weakness and failure in this race of bio-politics. In this way, COVID-19 not only serves as a prime incentive for the transmutation of the capitalist machine, but is furthermore a globalized declaration of the ascendancy of speed-Leviathanism: it perversely demands that all nations become a *dromocracy*, in the words of Paul VIRILIO, instead of a democracy. The success or failure of a nation is now wedded to dromology, rather than political science strategies in the traditional sense.[2]

But what's worth reflecting upon is this: must the progression of "cultural experimentation" be bound by the temporal logic of speed-Leviathanism, also restricting it to a narrow path of social acceleration? In fact, introducing a completely different process concept and developmental pace to the idea of "experimentation" is precisely what C-LAB has been demonstrating to the outside world. As a new type of cultural institution, it is under tremendous pressure from national leaders and public representatives to deliver visualizable "achievements of government" under the mentality of performance evaluations, as well as forces that push for the redevelopment of the site. In all fairness, to deal with these external demands (or to carve out a space of autonomy for internal experimental projects), C-LAB has recently begun organizing exhibitions in the same vein as other art venues, replete with what may be an excess of workshops, lectures, forums, and seminars, in a bid to increase its exposure or following, maintaining a continuous output for purposes of public perception.

Yet the hallmark of C-LAB, what distinguishes it as a "cultural experimentation hub," must be the unique design of the CREATORS Creation/Research Support Program (CREATORS Program). That is because it differs from other residency programs or subsidization platforms, in that the result or process evaluation isn't the sole aim, nor does it require that artists and artist groups deliver a clear "creative experimental outcome." Besides providing the requisite resources, spaces, hardware, and software facilities, the CREATORS Program focuses on agency between expert consultants, as well as the support of observer systems in establishing a foundation on which creative thought flourishes. In other words, the CREATORS Program's initial design is to remove two curses which have haunted contemporary art in Taiwan: one is the assumption that the experimentations and explorations of creative processes must culminate in the exhibition or production of a work. The second is the supposition that creative practice can always be exhibited or will always occur in the direction that leads to exhibitions and end products. Although it's called a "project" on the surface, in the increasingly bureaucratic world of art subsidies (with everyone scrambling to fill an endless stream of paperwork and progress reports) and thoroughly benchmarked programs of today (all subsidized/unsubsidized records must be presented in glanceable form), the CREATORS Program more or less preserves an air of unpredictability in artistic systems — that which cannot be assigned a rigid timetable, and does not readily agree with the ideas of typical project planning. It even permits certain segments of creative experimentation to remain unexhibited and unaccelerated. Put succinctly, the initiation of the CREATORS Program show that we still take stock in the time-consuming and resource-heavy demands of "cultural experimentation." Not only is it hard to obtain immediate results, but we must also acknowledge the fact that "experiments do fail," and with it bring a certain disruptive element to the exhibition-seeking, turnover-obsessed world of contemporary art.

Having said that, the motive is not to explain that "cultural experimentation" can only exist within a high-stakes, resource-intensive logic of art production. After all, we are far removed from the rawness and unreserved production scene of the Taipei International Post-industrial Arts Festival in the 1990s. On the contrary, the existence of CREATORS allows us to ponder: what are the extant flaws and blind spots of the unstoppable "great project plan era"? Could it be that we lack the capacity for long-term project planning? That we do not know how to set aside enough time for experiments, trials, explorations, and thought, and to allow creativity to age and evolve?

The non-transparency of projects and the myth of knowledge visualization

Of course, the CREATORS Program in its current form is not without its challenges. On the one hand, it faces pressures of resource allocation. On the other hand, the number and diversity of project types that fall under its jurisdiction is profound, spanning Bio-Art, new media art, sound art, scientific research, historical survey, and even regional revitalization. Strictly speaking, unless C-LAB's regular staff includes specialists from more diverse backgrounds, it will be hard-pressed to handle such a wide variety of projects. That is to say, if the CREATORS Program is limited to the role of a resource platform and intermediary (despite the difficulty of accomplishing even this), it can only be an art residency platform that is transitory and desultory. Despite a principled allocation of resources towards smaller projects, it cannot coalesce into a clear theme of "cultural experimentation"; it can potentially spark discourse on several fronts, but due to a lack of depth in perspective and long-term stewardship, it cannot easily form a cultural settlement. Here, "to settle" means allowing people to take root, to let the rhythm of creative life continue steadily, creating an environment for "making mistakes" and "trial runs." These conditions cannot be attained in a state of constant alternation and rotation.

Moreover, the core issue of the CREATORS Program is being unable to shed the specter of the "logic of visualization," consigned to the circumstances of "exhibiting in irregular spaces," in conflict with its original design.[3] This inevitability is fully reflected in the recent Signals: CREATORS 2020-21 exposition, as exemplified in the words of its curator YU Wei[4]:

[3] CHU Yian, "Reflecting upon C-LAB: An interview with LAI Hsiang-Ling, Director of Taiwan Contemporary Culture Lab," ARTouch website, accessed January 28, 2021.

[4] Excerpt from the *Signals: CREATORS 2020-21* exhibition brochure.

— The exposition as a vehicle for public declaration is swept up in the cultural system's desire for transparency — its exhibition allows projects to be measured and assessed. Cultural practice is anchored by the official rules of submission, made transparent under the gaze of the system, becoming an entry on the scorecard of collective national competitiveness. At the other end of the spectrum, independent culture producers enter a cycle of writing project proposals, application, execution, documentation, interim reportage, concluding review, and waiting for the final reimbursement. Yet the private life cycle of each project does not end there, stoked by a greater passion towards the topic on an alternative time scale. The more grandiose the project, the more difficult it is to conclude, if it can be concluded at all.

Apparently, the curators themselves understand that the reason "cultural experimentation" must be exhibited mostly boils down to the national arts and culture bureaucracy's demands for visualization. Thus, all research projects of varying speeds, or processes on different time scales, must fit within a uniform deadline, accompanied by a decorous accoutrement of research documents, working sketches, or working models to represent tangible and impressionable "results" (or the glorified experimental "node"). As the window for systemic support closes, almost none of the 16 resident art collectives can boldly defend their projects, betraying aspects of their work that can "neither be exhibited, nor be accelerated" — which is to say the aspect of artistic practice that is rigorous but all-consuming, or an aspect of "cultural experimentation" that is non-transparent. As a result, the Signals exposition may have temporarily satiated the outside world's curiosity, but the gaze which seeks to capture the creative experimentation process is in itself two-dimensional and instantaneous. The more one succumbs to this logic of visualization, the further the exhibit falls into an awkward state of fragmentation and segmented representation, making it difficult to consider the original context of each individual project. Even though the topic tackled by each artist collective is interesting, the space constraints at the exposition do not allow a coherent presentation of the disparate projects, distracting from one another so that the audience can hardly grasp the work or its potentiality through a glimmer of fragmented documents.

The key issue stated above is well-established and cannot be resolved by C-LAB alone. This is because the logic of visualization has roots that go back at least to the exhibition Taipei Biennial 2012: Modern Monsters / Death and Life of Fiction. Ever since the Biennial in 2012, not only has contemporary art in Taiwan shifted its gaze fervently towards historical documents, but has also outwardly assumed the visual arts exhibition style of "museum/museology," paving the way for heated discussions on the meanings of the word "field." Despite the significance of this change in creative direction, which continues to this day, its "side-effects" were also apparent: exhibitions have become by default

[5] For critique on "reproductive thinking", see WANG Sheng-Hung, "Field Research: an initial survey of methodologies of contemporary art practice," originally published in *Art-co Monthly* magazine issue 253, October 2013, pp. 114-117.

a high-speed interface and platform for knowledge exchange. Whether the curator or artist exhibits a completed work or the representation of its creative process through field research data, maps, and handwritten notes, the "immediate exchange of knowledge" has become the implicated premise of the exhibition. The more the exhibition invokes the museum format (which usually packs in vast amounts of information without reservation), the more it is assumed that the audience, after making a round of the floor space, can walk away with the complete set of artistic knowledge. This is the prototype for the logic of visualization.

Nevertheless, the assertion that everything can be transformed into seeable/visible/perceptible/interchangeable states often neglects the fact that in the "primary scene" of field research or experimental exploration, there are countless unseen "somatic knowledge" that exists. These may be conveyed through textual exposition for the audience, but cannot be readily imparted; they require contemplation, gestation, and realization over time, or even needs to be "lived in" if it is to be acquired. If contemporary art presumptuously holds that this slow knowledge, which must be taken in on a different procession of time at a different scale, can all be transposed into the "secondary scene" of the exhibit (with the tagline "making the unseeable seeable"), or even believes that the secondary scene can stand in for the primary scene, we may well fall unknowingly into a kind of expository violence of "reproductive thinking." [5]

All in all, if we have no awareness of the implicit assumption that "everything can be placed in the exhibit," then we will naturally abuse any manifestation of the "work in progress" and find it hard to resist a logic of visualization. But this immediate, quick-acting tendency not only faces difficulty in deepening knowledge production systems of contemporary art, but also suppresses the artist's imagining of the experimental project's time scale, subdued by an omnipresent expositional violence. Thus, not only does "cultural experimentation", cease to become our last bastion of reflection against speed-Leviathanism, but those of us stuck in the systemic transparency myth, along with the new ethics and methodological discourse afforded by "cultural experimentation", also has become the dying breed in an atmosphere which takes social acceleration for granted.

Reflecting on the internalized forms of "cultural experimentation"

Strictly speaking, the key issue here is not having to choose between the "exhibitable/not exhibitable," The real problem lies in figuring out: which part of "cultural experimentation" lends itself to exhibition? For example in the Signals exposition, the Engineering of Volcano Detonating exhibit consists of just a handful of articles and event photos, that fail to convey the young artist collective's perseverance in the "transversals" of geology, archeology, astronomy, or even study of UFOs, nor their ambitiousness in searching for the rifts and dialectical relationships between

Editor's note

In addition to displaying the archive in the gallery, both artists had launched performing events during the Signals exhibition as part of the exhibition's public program, including Wu Ping-Sheng's WFS system experiencing session, *Anteriusound Series – Soundcheck#1*, and TM LI's live audiovisual performance, *Echo/ Reflection I*.

science/non-science. The project's latent worth is much more than its images or objects which are collected in the exhibition space, but rather as their series lectures presents it, a kind of knowledge community shift that is hard to induce through existing disciplinary frameworks. Let's take WU Ping Sheng's *Transonic 2020* and TM LI's *Freeze Historical Reverb: the Perceptual Narratives in the Caves in Taiwan* as two further examples. The former has only the working sketches and speaker arrays to show, which despite being subjected to repeated experimentation remain prototypes. The audience can feel and understand the "visual aspects" of the installation and techniques, but not necessarily appreciate its operational principles through the audio space provided by C-LAB. Similarly, the latter exhibition provides a lavishly adorned listening room of historical objects, which alas provide only an approximation of a "near-live situation," failing to supply enough information to help the visitor understand the technical thinking behind Ambisonics and tunnel recordings, not to mention the connection between the technology itself and Taiwan's mining industry or military heritage sites. That is to say, both WU Ping Sheng's and TM LI's respective projects have the potential to be adequately exhibited, just not in their current states. The current displays of installation and experiential representation may have satisfied the cultural system's desire for transparency over the experimentation process, but are bound utterly by a uniform knowledge projection of the progression, hastily trying to adopt a knowledge format that the visitor instantly picks up on. Editor's note

It bears saying explicitly, that the above problem cannot be relegated to a fault in the exhibition's presentation. The main reason is that these projects are variously restricted by a logic of visualization, so that the artists are required to either prematurely adopt exhibitory thinking in the experimentation phase, or forced to reveal its "primary scene," exposing the non-transparent aspects of the "cultural experiment." In other words, the key focus here is not whether "cultural experimentation" can assert its opaqueness or not, in order to recede from public scrutiny and inspection.[6] It may well be that we have misconstrued the core task of <u>"cultural experimentation", that rather than manifestation/demonstration, it is a store, accumulation, and reserve for creative energy</u>. Here, the true essence of the term "experimentation" is to find a workable implication format towards creative thinking, allowing its energy to converge and coalesce, to be unleashed when the appropriate time comes. Any institution that shares this core value should never have to expend its precious energy on ceaseless cycles of exposition production.

As the above diagram shows through the scheme of plant life, the typical "cultural exhibition" is like a grapevine, taking away productivity and nutrients from the soil. The surrounding solar, air, and water conditions determine how it prospers. The essence of the "cultural exhibition" is to absorb the cache of accumulated cultural energy, allowing creators' notions to be realized through the actualization of creative thinking. Its basic premise is stratified planning, allowing prearranged artistic content to develop one at a time,

[6] That is to say, the issue lies not in WU Muching's observations of the arts and culture circles' habitual "disdain towards being managed". Not that it's untrue, but the issue is an inability among artists to strike a balance between the cultural system's demands for transparency and the present paces of creative experiments. See: WU Muching, "The demise of C-LAB's empty plots, the Goonies of the arts and culture circles, originally published" in *Artco Monthly* magazine issue 337, October 2020, pp. 102-103.

resulting in a tangible and palpable final result. In comparison, "cultural experimentation" is like a grape seed that stores energy and genetic sequences, motivated by goal of storing the multiple energies of creative life within the one seed, for the purpose of producing an embryonic origin for creative life, an enveloping movement. The key to "cultural experimentation" is not to grow as many variants of the grapevine as it can, but to find a suitable format that can best preserve the pure potentiality and virtuality of artistic practice. Then, if "cultural experimentation" must be outwardly displayed, it can be at most a form of interiorization through which the "many" could be glimpsed through the "one" — these could be an informal exchange of numerous intense but nonspecific results, or it can be the building of thought-provoking relationships without a definitive time frame — the key here is to distance oneself from the instantiative logic that "everything can be added to the exhibit," and gradually form a normalized community with interiorization as its core enterprise. Put simply, it is an attempt to promote a cultural ethos. As in the original intent of the CREATORS Program, it is a constant allowance of keeping a certain "distance" from reality (yet always being in touch), in no hurry for realization, without rushing towards exposition, constantly holding oneself to a state of potentiality, and gradually piecing together a sensible "spiritual corner" in which creators and researchers can produce slowly and think slowly.

Today, as each major cultural institution churns out exhibitions at a breakneck pace, each art school engrossed in metrics of "instructional excellence," "vision projects," and "innovation," dare anyone proclaim their imperviousness to the onslaught of speed-Leviathanism, adherent to a program of virtualization by focusing on the accumulation of cultural strength? This was one of the most unique qualities C-LAB has to show. But nowadays, swept up in the storm of expedited production, it may no longer be able to reflect upon the rush of the document exhibition format; it has been brought to the level of a contemporary art exhibition center like any other.

Honestly speaking, the more exhibitions C-LAB produces, the more it diverges from the original intent of the CREATORS Program. This is because any exhibition takes up vast sums of investment in human resources and focus. Despite the sufficiently large operational crew of C-LAB, as long as it continues to approach "cultural experimentation" from a thinking of actualization, then its frontline staff must face the pressures of presentation, working with the inefficiencies of fragmentary spaces to rival the number of exhibitions a typical arts and culture institution can produce. At the same time, they must do so under extremely compressed time frames, providing all manners of support activities in hosting, hiring, and stewardship. Burning the candle at both ends, this becomes a serious drain of energy on even the most well-organized of teams, and is not the way to cultivate and establish a firm foundation. At the same time, C-LAB is pressed to "reinvigorate the old with the new," creating a false understanding and expectation among the public, who has come to understand and experience the shallower aspects of "cultural experimentation" on an inappropriate time scale and exposition schedule. Thus encumbered, C-LAB undoubtedly suffers in an overworked state.

It is an unshakable curse of Taiwanese contemporary art that assumes artistic practice must walk a "solid line" (that is, creative endeavors cannot follow an unrealistic "dotted line"). However, in the ingrained belief that everything must be published, supervised, tracked, and written off by the arts and culture bureaucracy, the dilemma between actualization and virtualization is first and foremost C-LAB's burden to bear. This then may be the most formidable challenge C-LAB faces today.

Author's Bio

WANG Sheng-Hung

Having earned a doctorate from the School of Fine Arts of Taipei National University of the Arts, Wang currently works as an assistant professor at the Graduate Institute of Art Studies, National Central University. His main research interests include contemporary video art, performance art, and body art, with a focus on contemporary biopolitics and their impact on physical action and mode of expenditure under the prevailing accelerationist politics and bureaucratic apparatus. On this foundation, he reflects on the production and expansion of local art theories in Taiwan.

• References

1 Website of the Engineering of Volcano Detonating: https://spacestudio1219.wordpress.com/

2 SHEN Ke-Yu, Not at this time: the potentiality of time differences, CLABO website, March 8, 2021.

3 FENG Ping, The sampling, *recording, and reconstruction of sonic landscapes*, CLABO website, March 2, 2021.

4 WANG Sheng-Hung, The state of art production in times of bureaucracy—discussions on the curatorial statement and consciousness of the exhibition *Are we working too much?* Part I, CLABO website, October 13, 2019.

5 WANG Sheng-Hung, The realm of art production in times of bureaucracy—discussions on the curatorial statement and consciousness of the exhibition *Are we working too much?* Part II, CLABO website, October 13, 2019.

2018-2020 CREATORS 回顧討論會

整理＿林怡秀

臺灣當代文化實驗場（以下簡稱 C-LAB）CREATORS 計畫，自 2018 年開始至今已邁入第四年，除了透過徵件評選的「創作／研發支持計畫」主體，也逐年慢慢建立起「陪伴觀察員／年度觀察員」制度，期望能在以往對於計畫的成果期待之外，更深化實踐與實驗的過程意義。2020 年更透過「年度觀察團」的設計，提供計畫觀察任務更具主題企畫的向度，以及後設視野的討論可能。此次的回顧討論會，主題關注於：CREATORS 計畫的核心精神、如何在過程中產生公共意義？「計畫觀察」這項任務帶有怎樣的意義？以及，這項計畫的執行有哪些矛盾與困境？

時間｜2021.09.27（一）14:00
主持｜游崴（2021 CREATORS 計畫統籌）
與談｜魏妏潔（2019-2020 CREATORS 計畫統籌）、王聖閎（2018、2021 陪伴觀察員／2020 年度觀察團顧問）、謝鎮逸（2019-2021 年度觀察員）、馮馨（2018 陪伴觀察員／2020、2021 年度觀察員）、林怡秀（2019 進駐團隊成員／2020 年度觀察團總召）

游崴（以下簡稱游）：CREATORS 計畫從 C-LAB 最初啟動就開始，很大程度上也成為 C-LAB 定義自身文化實驗想像及核心精神的元素，與其他藝文補助不同，CREATORS 是「支持」計畫而非「補助」計畫，重心放在發展過程中的公眾性而非最終成果，也希望藉由每年評選的過程、徵選的廣度去定義「文化實驗」是什麼。CREATORS 計畫進駐時間與評選方式隨每年的狀況進行調整，2019 年開始，CREATORS 分為創研進駐與創研支持，對外發表的部分除了 2018 年、2020 年的成果展，主要還是開放工作室，2021 年因為疫情，開放工作室改為線上舉行。觀察員制度則從第二年起分成陪伴觀察員、年度觀察員兩種，前者定位在顧問性質，後者著重書寫。自第一屆開始就與《CLABO 實驗波》合作，2020 年開始改為觀察團。今天的圓桌討論，主要想談對觀察員角色的建議，以及對 CREATORS 機制的觀察。首先請妏潔分享你 2019-2020 擔任統籌時的想法。

魏妏潔（以下簡稱魏）：最初大家對於觀察員的想像比較偏向專業諮詢，但第一年有些觀察員不一定能完成文字書寫的要求，跟執行長討論後覺得應該諮詢、書寫兩者分開，所以 2019 年就用類別區分，當時也有一個狀況是三位年度觀察員的文字風格、撰寫的方向有一定的落差。2020 年就有比較明確的統籌規劃，有觀察員課程、工作坊的協助，這也是 CREATORS 計畫中延伸出很可貴的東西，不是只有進駐團隊在做實驗創新，CREATORS 團隊本身也在實驗。觀察員歷程的實驗其實很有趣也很必須，這成為計畫滾動調整時一個蠻有趣的現象，我們也在找好的模式去觀察這些團隊，這些觀察會更貼近觀眾，內外都有兼顧到，這是在觀察員上比較明確的調整跟實驗。

林怡秀（以下簡稱林）：我第一年是觀察員、第二年進駐、第三年是觀察團總召，角色的變化剛好都有感受到 CREATORS 在不同階段的需求，2020 年一開始很單純是《CLABO 實驗波》需要一個編輯彙整文章，剛好我近幾年有積極想和年輕作者合作，CREATORS 提出的補助金額、觀察員稿費其實是很好的機會，我當時想的是，可以找不同領域的年輕作者進來這個觀察過程，所以這一年找來包括生物科技、表演藝術、視覺藝術，甚至歷史背景的人。也因為大家並不是都在藝術圈工作，所以有做書寫工作坊，討論怎麼做觀察寫作、分享經驗，分享經驗的過程還包括當時的現象、展覽、書寫方法的交流，這一年除了進駐團隊，觀察員本身也是一個實驗的過程。

馮馨（以下簡稱馮）：我是 2018 年的陪伴觀察員，當時的狀態其實很疏離，只能透過探訪知道藝術家的進駐狀況，但是 2020 年的邀請非常不同，那樣的討論狀態是少有的。可以透過觀察員們在工作坊的分享，知道其他團隊的計畫進展，加上成員組成差異很大，工作坊並不是用當代藝術的邏輯在看這件事，而是打開另一個文化討論的向度，關注相對很本質性的問題。當時我們很積極討論「什麼是文化實驗？」、「觀察員的角色是什麼？」，三個階段的工作坊討論都聚焦不同事情，並且延續到大家下一次觀察階段，這個進展是很明確的。雖然不確定藝術家彼此之間的交流如何，但觀察員彼此的討論狀態很熱絡。那三場工作坊對我是很重要的書寫交流，可以在文章刊登前預先看到大家的書寫狀態，使我無論在書寫或是討論上，都可以來回推進思考書寫者之於藝術家與機構的關係，是蠻好的經驗。

謝鎮逸（以下簡稱謝）：我是 2019 年到現在連續三年當觀察員，現在的我會先確定團隊如何看待觀察員，這會很大程度決定觀察員接下來的書寫行動。當觀察員提出要探班、觀察團隊工作時我會遞進的探訪，如果藝術家可以聊，才會進一步討論。所以我要先拿捏跟團隊的距離跟關係，觀察員要先取得藝術家的信任，藝術家才會願意多聊。觀察員的位階有時候也會影響團隊想不想跟觀察員溝通，這三年來可能因為團隊不夠了解我，會覺得我只是報導記者，或覺得這只是計畫中一個需要完成的任務，這些態度就會影響到交流品質。另一方面，也要考慮觀察書寫到底要不要設定讀者？怎麼拿捏觀察的深淺度？如果單純講這次計畫到底應該談什麼？因為一方面他們的實驗還沒有作品化，我們要基於什麼基礎去生產內容？

王聖閎（以下簡稱王）：我先簡單兩點回應，剛剛聽下來初步感想是 C-LAB 如何創造這個專業社群裡年輕作者、藝術家的相互觀看或交流，對提升這塊的品質很重要，我們應該優先設定一些交流形式，不是要對外、對立委、對文化部，而是服務專業社群中不同位階、組織架構背景的人，讓他們彼此觀看、討論，一旦這件事有做到，工作的人活力就會出來，也會減緩面對外部的焦慮。上次的工作坊也讓我感受到它難在一方面要創造不同領域適當的交集，同時又要凸顯差異。也許我們可以把工作重點放在這，形成內部觀看、交流的品質會是很重要的指標，也許先以專業社群為主，假使未來機制成熟，觀察員的機制還可以往前推一步。2021 年於 C-LAB 舉辦的展覽「未來相談室」，它提供一個特別的對話介面，可以把適當的對話機制推進到不同專業與學科訓練背景的交流關係裡，我們是否可以把觀察員制度跟這樣的對話機制結合起來？

魏：這個計畫珍貴的地方是有很大的包容性，因為我們沒有成果發表，行政團隊只能靠經驗去判斷什麼時候要進入作品，什麼時候要退出來，這個拿捏是藝術進駐會遇到的狀況，團隊狀況百百種，這時候做藝術行政的同事就很需要觀察員的專業協助，要去碰撞才能找到與藝術家交流的頻率。我們不會去涉及計畫內容或製作，我們是在計畫上去做協助。不論藝術家本身或是行政本身，這邊還有一個重點就是：到底要逼藝術家逼多緊，要看到多少成果但還能保有一個自由性、彈性去實驗？我自己也還沒想到明確的方式去認定怎樣的程度可以稱為成果，到底底限是什麼？團隊自己要看到什麼？有沒有辦法被認可？像 CREATORS 這樣的計畫，是非常可貴的，但這個計畫很有可能某天會消失，因為現在的補助結構還是 KPI 導向，還沒有到這樣的高度去認同跟支持。

王：在臺灣目前的環境，過往建立的系統本身有個很強的教育藝術家的傾向，那是一個「主體性的形塑工程」，這是大環境的特性。但在這樣的環境中，CREATORS 要展現一種對抗，一方面要對抗被補助系統完善的工程，一方面要提出另一種方案去批判這些固定模版的主體性工程，也許在徵件上就可以先去說明這不是 KPI 型的計畫。

游：我覺得要打開過程中的公共性，通常是靠文化生產鏈的中間人角色，例如編輯、教育推廣者，他們有點像是在處理第二現場，可能是工作坊、討論這類活動，這些活動本身會變成計畫不可或缺的部分，在很多計畫型創作裡，藝術家也會變成中間角色，像是把第一現場的訪談轉譯成可以利用的資源，使這個資源本身具有公共性。

魏：要讓藝術家知道 CREATORS 的精神，通常是最一開始的說明會，我們本來想像讓團隊自己去互相交流，但其實是很困難的，透過分享會是一種方式，但對藝術家來說也會有負擔，尤其是這種醞釀型的計畫。2020 年我們有試著辦小聚會，促進彼此的關係，但如果要很深入了解彼此的東西就需要一點強制性，不然很難自然發生，這個都是可以在行政端去設計的。公眾性對我來說是跟不同領域的小眾有關，無論是粉絲頁宣傳、意見領袖，散播出去都是對一個個小眾，像這樣的討論，如果要求藝術家都要被大眾看見很難，所以這個公眾性一定要多元。

謝：但 C-LAB 的期待不一定符合團隊現階段的需求，因為回到藝術家團隊自身，他們有沒有想要公共化？以及，藝術家有沒有準備好要展示這些過程？因為其中會有很多失敗、嘗試，這些有要被公開嗎？沒有要求結果的話，藝術家跟我們並肩同行的是什麼？是發展過程，或是推廣活動、講座？有些團隊可能是把這邊當成創作生涯的小驛站，所以黏著度不高，進駐密度也不高，他們把這邊當成驛站，然後繼續走自己的下一步。另外是觀察員的書寫，我們很難確認訪談是否有效，觀察期程的中段就定稿的危險是這些內容未必是藝術家的概念，這個觀察到最後要因應團隊進度，過程中的觀察報告很可能變成只是針對過程的報告，或是觀察員自己的擴充評論，矛盾點在於作品還沒有明確結果論，但報告要不要有結果論？如果書寫沒有可以掌握的基礎或現象、事件，評論可以怎麼寫？如果我們交出的文章資訊破碎，讀者要了解的到底是什麼？這個計畫立意良善，但最後可能連觀察員的角色都很模糊，有經驗的書寫者可能還是可以做文章，但回到公共化的問題，觀察員到底能不能幫藝術家講話，這是我還在思考的問題。

王：就我的觀察，過去不管是補助或是美術館獎項，他們大部份都有一個傾向是讓這個機制本身盡量不透明，因為機制凸顯出性格或評論趨向就會被攻擊或針對，所以都是代言制，它本身不說話，但邀請評委、觀察員講話，這些系統沒有一個「化身」（avatar）讓外界知道這個機制在想什麼。順著這個，CREATORS 是否需要一個化身？例如陪伴機制本身的特性，這樣的化身會不會起到溝通效果，它會跟傳統系統有什麼不同？現在 CREATORS 比較沒有這個角色，這個傾向還是模糊的，還是有一個距離，我覺得也許可以設想這個計畫本身的書寫機制、陪伴機制的化身是什麼狀態與性格，讓對話更聚焦。CREATORS 觀察員其實也需要跟 C-LAB 對話，目前缺少一個可以持續跟觀察員、行政團隊對話的橫向層次。如果有種化身機制能夠反覆闡述計畫的核心精神，寫作者就不會只是跟藝術家、公眾去對話，這個對話就會有一種思想推進的效果。

馮：我第一年很快就設定了我的定位，因為我觀察的對象幾乎都是小眾領域，引介跟說明是首要目標，所以書寫這些聲音藝術領域的 CREATORS 時，我都會處理脈絡介紹跟技術說明的問題，因為讀者是未知的一般群眾，雖然實驗很重要，藝術也需要時間醞釀，但我更希望讀者可以看到這些 CREATORS 怎麼開展這個創作與實驗的過程。我的書寫沒有太多糾結，不是為了藝術家或是機構，而是為了讓大眾可以認識創作的過程。當初若沒有這個書寫定位的預設，其實是會迷失的。

游：接下來也想問大家心目中認為一個「理想的」文化實驗計畫應該要有的模樣？

林：「藝術」可能不是這個問題的關鍵，但 C-LAB 所有計畫、活動最後都是用藝術的方式去表現過程，CREATORS 裡像是林傳凱這種團隊，選擇用藝術的方式去呈現他長久以來收集的田調過程，因為這種類藝術的呈現方式可以去觸及或是推展這個題目。CREATORS 這個補助跟支持計畫很好的是沒有限制來的人是誰，也不限制最後的產出。我想起有一次工作坊時觀察員王順德分享：「我們這些進實驗室的人，其實並不知道我們要實驗的是什麼，這個結果是沒有想像到的，這個過程才是實驗，因為如果一開始就知道結果，那就不是實驗了。」你必須不知道結果，實驗的動作才會有意義，這就是 CREATORS 有趣的地方，它允許失敗，或是實驗出不一樣的東西，理想的文化實驗應該是你甚至不知道結果，只要告訴我你想做怎樣的實驗就好，最後有沒有產出就是另一件事。

謝：我們都知道不要求結果的計畫多麼被環境所需要，但有個可惜的地方是，這個點沒有被好好闡述，所以機制上的設計應該想辦法讓這個立意不要被消化成驛站的形式，讓計畫的理想性可以被確實的定調、被好好的使用、往更大的想像去發展。如果要形塑出文化力量或實驗力量，C-LAB 能否讓內外部的能量持續跟書寫社群串連，因為如果沒有持續的社群或機制將會很難做到，時間量度會是很大的問題，很多計畫不是 6 個月就可以完成，團隊要怎麼保持計畫給他們的能量，書寫者可以怎樣更了解、面對機制的討論？以及這個計畫帶給我們的持續性到底可以是什麼？

王：現階段臺灣的大環境，幾乎所有藝術家都在想怎麼回應補助系統，想盡辦法在計畫書展現自己，大家都很緊繃，在這個制度下所有創造性都是被逼出來的，但 CREATORS 這邊的確給出一個不同的思路，剛剛談「驛站」的狀態也許不是壞事，只是他們把驛站的狀態的可能性想得很小：因為接下來還有很多要硬拚的計畫，所以我在這邊休息一下。我們是否能夠讓藝術家知道有一種創造性恰恰是要在這種修復式的、休息式的、喘息式的、調整步伐式的實驗狀態下自然產出的？目前的大環境並沒有教育藝術家有種創造性是在另一種放鬆狀態下產生出來，而是讓藝術家以為都要有漂亮、可見的成果。也許 CREATORS 可以試著調整，讓藝術家知道不單單只是把這裡當成驛站，而是在這樣的計畫模式下可以有不同的生產性。我期待這個計畫可以提供一些教育取向，也許可以改變臺灣現在太過同質化的創作狀態。

2018-2020 CREATORS Review and Discussion Meeting

■ Compiled __ LIN Yi-Hsiu

The CREATORS Creation/Research Support Program (CREATORS Program) launched by the Taiwan Contemporary Culture Lab (C-LAB) since 2018 has now entered its fourth year. In addition to the open call for "creation and research residency/support" as its main component, the program has gradually established an "Consulting Observer/Reviewing Observer" system over the years in hope of deepening the meaning of the processes of such practice and experiment apart from the usual expectation of the program's results. Moreover, the design of the "Reviewing Observer Collective" started from 2020 has unveiled a more thematically planned aspect as well as a possibility for discussions from a meta-perspective in terms of the observation mission given to the program. This closed-door discussion meeting focuses on various aspects: The core spirit of the CREATORS Program, how to generate its public meaning during the processes, the meaning of the mission that is known as "program observation," as well as the contradictories and difficulties encountered throughout implementing the program.

— Time | 2021.09.27 (Mon.), at 14:00
Moderator | YU Wei (2021 CREATORS Program Coordinator)
Discussants | WEI Wen-Chieh (2021-2020 CREATORS Program Coordinator), WANG Sheng-Hung (2018, 2021 Consulting Observer; 2020 Reviewing Observer Collective Consultant), Yizai SEAH (2019-2021 Reviewing Observer), FENG Hsin (2018 Consulting Observer; 2020, 2021 Reviewing Observer), LIN Yi-Hsiu (Member of Creation/Research Residency Group in 2019; 2020 Reviewing Observer Collective Convenor)

YU Wei [referred as YU below]: The CREATORS Program was launched since the inauguration of C-LAB, so to a considerable degree it has become a component that defines the imagination and core spirit of C-LAB's cultural experimentation. This makes it very much different from other arts and cultural grants. The CREATORS Program aims to "support" rather than to "subsidize," and its emphasis is on the publicness that emerges in developmental processes rather than final results. Also, it is set out to define "cultural experimentation" through the process of annual selection, as well as the expansiveness of the open call. The time of residency and selection of the CREATORS Program have been adjusted annually based on the circumstances of each year. Since 2019 onward, the CREATOR Program has divided into "Creation/Research Residency" and "Creation/Research Support." Apart from the public final presentations of 2018 and 2020, the program focuses primarily on open studio. In 2021, because of the pandemic, an online open studio was held instead. Since the second year of the program, the observer system has divided into the "Consulting Observer" and the "Reviewing Observer," with the former focusing on providing consultation and the latter reviewing the processes of creation. In 2020, the system of an observation group was adopted, which combined annual observation. This roundtable discussion today aims mainly to discuss the role of the observer and suggestions for it, as well as your observations of the mechanisms of the CREATORS Program. To begin, let's hear Wen-Chieh's thoughts on being the program coordinator from 2019 to 2020.

WEI Wen-Chieh (referred as WEI below): At first, everyone's imagination of the role of the observer tends to be related to professional consultation. However, in the first year, some of the observers could not fulfill the requirement of writing. After discussing with the CEO, we came to the conclusion that providing consultation and writing should be separated. So, we separated the two in 2019. Bach then, one situation we encountered was that the style and direction of writing of the three Reviewing Observers varied to a certain degree. In 2020, we had more specific coordination and planning, namely the assistance of the observer course and workshops, which are also valuable outcomes produced by the CREATORS Program. That is to say, not only the residency groups are looking for experimental innovation, but the team running the CREATORS Program is also experimenting. The experimental journey of the observer system has been interesting and essential. This creates a rather intriguing phenomenon when rolling-wave adjustments are made for the program because we have also been searching for a good mode to observe the residency groups, and observations derived thereof can be closer to the audience. So, both the internal aspect and the external aspect are considered . These are the specific adjustments and experiment that have been made in terms of the observer.

LIN Yi-Hsiu (referred as LIN below): I was an observer in the first year, a member of a residency group in the second year, and the observation committee group convenor in the third year. So, the change of roles has enabled me to perceive the various needs of the CREATORS Program at different stages. In 2020, it was because *CLABO* needed an editor to gather and organize articles, and I happened to be thinking about working more actively with young writers in recent years. With the amount of the subsidy and the renumeration for observers provided by the CREATORS Program, it was a rather perfect opportunity. At the time, I was thinking that I could find young writers from different fields to participate in the observation process. So, over the past year, I invited writers from the fields of biotechnology, performing arts, visual arts, and even history. Because not all of them work in the art circle, writing workshops were organized for discussing the methods of observation writing and for sharing experiences. The process of experience sharing also included exchanging ideas about the phenomena, exhibitions, and writing methods at the time. So, not only the residency groups from that year but also the observers themselves constituted an experiment.

FENG Hsin [referred as FENG below]: I was an accompanying observer in 2018. The situation at that time was quite aloof. I could only get to know the situations of the artists-in-residence through visiting. Yet, it was very different in 2020. The discussions we had were very rare. We could know about the projects of other artists through other observers' sharing made in the workshops. Also, the members [of the observation group] were very different, so the workshops did not operate on the logic of contemporary art but rather opened up another dimension of cultural discussion to focus on some relatively intrinsic questions. At that time, we discussed very actively about questions such as "What is cultural experiment?" and "What is the role of the observer?" The workshop discussions in three stages concentrated on different subjects, which were carried into the observation in the next stage. So, the progresses were very clear. Although I could not be certain about the exchange among the artists, the observers had very lively discussions. To me, the three workshops were crucial exchanges about writing, and allowed me to get to understand others' writing condition before the articles were published. Therefore, in terms of both writing and discussions, I could repeatedly think about the relationship among the writer, the artists and the institution. It was a very good experience.

Yizai SEAH [referred as SEAH below]: I have been an observer for three consecutive years since 2019. Now, I will first make sure that I know how the residency groups view the observer, as this determines the observer's subsequent action of writing to a very large extent. When I ask to visit and observe a residency group, I will visit the group step by step. If I can establish conversations with the artist, I will then take the discussion further. So, in my case, I will first evaluate the distance between myself and the residency group. The observer must first gain the trust of the artists for them to be willing to talk more. The position of the observer sometimes also affects the residency groups when it comes to their willingness to communicate. Over the past three years, perhaps because the residency groups did not understand me enough, they would see me as a journalist, or feel that this was simply a mission to be completed in the program. Such attitude would affect the quality of our exchange. Another thing to be considered is whether a readership should be preconceived for the observation writing. How can the depth of the observation be determined? If the writing is to simply talk about the projects, what should be included then? Because their experiments have not yet been converted into works, what should we use as the basis to produce the content of the writing?

WANG Sheng-Hung [referred as WANG below]: I shall first respond to two things. After what I have just heard, my initial thought is that it is very important in terms of how C-LAB facilitate and enhance the quality of viewing or exchange between young writers and artists in this professional community. We should first set up certain forms of exchange, which are not meant for the external world, for the legislators, and the Ministry of Culture, but for serving people of different positions and organizational backgrounds in the professional community to facilitate viewing and discussions between them. As long as this goal is achieved, the vitality of those working on this will be induced, and anxiety about facing the external world will be eased. The workshop last time also allowed me to perceive its challenge: on the one hand, it needed to create suitable intersections among different disciplines, and on the other hand, it had to highlight differences. Perhaps we can stress this in what we are doing. The formation of internal viewing and the quality of exchange will be a very important indicator. Maybe we can focus on the professional community first, and if the mechanism matures in the future, the mechanism of the observer could then advance. My inspiration comes from C-LAB's exhibition Your Future Now in 2021 which provides a unique dialogue interface that is able to incorporate a suitable dialogue mechanism into the exchange among different professions and academic backgrounds. Is it possible for us to combine the observer system and this mechanism of dialogue?

WEI: What makes this program precious is its vast inclusiveness. Because we do not have final presentations, the administrative team can only rely on experiences to decide when to engage in the projects and when to back out. How to achieve this balance is a situation one encounters in art residency. The residency groups could present a wide range of situations, in which our colleagues in the art administration department will need professional assistance from the observers, who have to collide with the residency groups to find the balance in their exchange with the artists. We do not interfere with the content or production of their projects. Instead, we provide assistance for their projects. There is another point for both the artists and the administration department: How much pressure should be put on the artists? How much result are we expecting while being able to provide a certain degree of freedom and flexibility for experimenting? I have not yet thought of any specific methods to define what degree of completion can be viewed as results and the minimum degree of completion for that matter. What do the residency groups hope to see? Are there ways to gain recognition? A program like CREATORS is extremely precious. However, such a program might disappear someday because the structure of subsidization is very much KPI-oriented and still lacks an equal height to extend its recognition and support.

WANG: In the current environment of Taiwan, the system established in the past shows a very strong tendency towards educating artists, which indicates a "formative engineering of subjectivity." This is the characteristics of the overall environment. However, in such an environment, the CREATORS Program demonstrates a form of resistance: while resisting the engineering perfected by the system of subsidization, it also needs to put forth an alternative to critique these regular modes utilized in such subjectivity engineering. Perhaps, the announcement of the open call can explain first that the program is not KPI-oriented.

YU: To me, to uncover the publicness throughout the process usually relies on the role of the mediator in cultural production, for instance, editors and education advocators. In a way, they are working with the second site, which can be activities like workshops and discussions. These activities will become an indispensable part of the projects. In many project-based creations, artists will also serve as the mediator who, for example, translates first-hand interviews into useful resource, consequently bringing out the publicness from this resource.

WEI: Explaining the spirit of the CREATORS Program to artists is usually done at the orientation. Our initial imagination was that the residency groups could carry out the exchange themselves, but this was in fact a rather difficult thing to do. Sharing session is one of the approaches, but it also creates pressure on the artists, especially for those working on projects that take time to develop. In 2020, we organized small gatherings to bring artists closer. However, for them to understand each other more deeply, these functions need to be somewhat compulsory. Without being so, it is difficult for such things to happen naturally. These activities can be designed on the administrative end. To me, publicness is related to minority groups in different fields. Whether it is through the promotion on fan pages or key opinion leaders, the communication is to reach individuals in minority groups. In discussions like this, it is very difficult to ask all artists to be seen by the public. So, this publicness must be diverse.

SEAH: Yet, C-LAB's expectation does not necessarily match the need of the residency groups at the current stage. Because, as far as the art groups are concerned, do they want to be publicized? Moreover, are the artists prepared to show the processes? Since there will be many failures and tries along the way, are these to be shown to the public? If it is not the result that we are seeking, what is it that the artists are working for with us? Is it the developmental processes, or the promotional activities and talks? Some residency groups might see this place as a relay in their creative careers, so their stickiness is low and their frequency of coming and staying here is low as well. To them, this place is like a relay, after which they will carry on and take the next step. Another thing is the writing of the observer. It is difficult for us to ascertain if the interviews are effective. The stake of finalizing the articles at mid-term is that the contents do not necessarily reflect the artists' ideas. In the end, the observations should correspond to the progresses made by the residency groups, so the articles might just be reports on the processes, or the observers' expanded reviews. A point of contradiction lies in the fact that there has yet to be any specific results regarding the works. So, should the reports focus on results? If there is no basis, phenomenon, or event for the writing, how else can the reviews be written? If our articles only include fragmentary information, what is it that the readers should understand? This project is founded on a good intention, but even the role of the observer can become blurry in the end. Experienced writers might possibly be able to come up with something; but circling back to the topic of publicness, I am still thinking about the question of whether the observer can speak for the artists.

WANG: Based on my observation, no matter it is a grant or an art prize given by a museum, a large part of it tend to retain a mechanism that is non-transparent. The reason is that if the mechanism shows a specific character or review tendency, there is a chance of it being attacked or targeted. So, a representation system is adopted instead. That is, the mechanism remains silent, but jurors or observers are invited to comment. These systems themselves do not have an "avatar" to help the outside world understand what they are thinking. Following this observation, does the CREATORS Program need its own avatar? For example, can an avatar, with the characteristics of the accompanying mechanism, achieve the effect of communication? How does it differ from the conventional system? At the moment, such a role does not exist in the CREATORS Program, and this tendency is rather indistinct. There is still a certain gap. I feel that maybe it can be helpful to envision the state and character of such an avatar for the writing and accompanying mechanisms to give the dialogue a clearer focus. As a matter of fact, the observers in the CREATORS Program need to dialogue with C-LAB as well. Right now, such a horizontal dialogue between the observers and the administrative team does not exist. If there were an avatar to repeatedly elaborate on the core spirit of the program, the writers would not be only dialoguing with the artists and the public. This way, the dialogue would have an effect of propelling the thinking.

FENG: In my first year, I had positioned myself rather quickly. Since the subjects of my observation were fields of minorities, introduction and explanation became my primary goal. So, when I wrote about these creators in the field of sound art, I would work on introducing the context and technical explanations because the readers were the unknown general public. Although experimentation was important, and art did need time to develop, I still hoped that the readers could see how the CREATORS develop their creations and their experiments. I did not have too much trouble with my writing. I did not do it for the artists or the institution but instead for the public to know more about the creative process. If I did not position my writing like this at the beginning, it would be easy to become lost.

YU: Next, I would also like to ask everyone the following question: What do you consider a “good” project of cultural experiment?

LIN: "Art" is probably not the key to this question, but all programs and activities of C-LAB employ artistic forms to present their processes in the end. In the CREATORS Program, art groups like that of LIN Chuan-Kai can choose to demonstrate the processes of their long-term field study in the form of art because the chosen art form enable them to engage in or further develop their topics. The CREATORS Program, as a subsidization and support program, is very good because it does not limit the qualifications of the participants nor the final outcome. This reminds me of what the observer Sirius Wang shared in a workshop. He said, "people like me who work in scientific labs do not really know what our experiments are. The results are not imaginable. It is the process that constitutes the experiment. Because if you know what the results are, it will not be an experiment." So, the results must remain unknown so that the action of experimentation can be meaningful. This is what makes the CREATORS Program interesting — it allows failures, or the possibility of creating something different out of the experiments. An ideal cultural experiment is that you do not even know the results, and you can simply tell me what kind of an experiment you want to conduct. As for the final outcome, that is another matter.

SEAH: We all know how much this environment needs programs that do not seek results, but it is a pity that this point has never been fully elaborated. So, the mechanism should be designed in a way that this conception is not reduced to simply providing a relay so that the ideality of the program can be affirmed, put to a good use, and develop towards a greater imagination. If C-LAB wants to shape a cultural or experimental force, will it be able to sustain both its internal and external energy, and to connect with the writing community? This is important because without a continuous community or mechanism, this will be difficult to achieve. The scale of time will present a huge problem. Many projects simply cannot be completed in six months. How do the residency groups sustain the energy provided by the program? How can the writers better understand and cope with the discussions about the mechanism? What can the continuation of the program bring?

WANG: In the overall environment of Taiwan at the current stage, almost all artists are thinking about how to respond to the system of subsidization, and are trying their best to display themselves in project proposals. Everyone is very tense. In such a system, all creativities are somehow forced. On the other hand, the CREATORS Program offers a different way of thinking. The situation of being a "relay" is not necessarily a bad thing. It is just that the artists might see the possibility of this "relay" in a limited way — because they have many more challenging projects to come, they will take a break here. Is there a way for us to inform the artists that a certain type of creativity is to be produced naturally in such a restorative, restful, respite-like, and pace-adjusting experimental state? The overall environment today does not educate artists about such creativity that is to be produced in a relaxing state, but rather makes artists think that they should all present appealing, visible results. Perhaps the CREATORS Program can try to make adjustments so that artists will not view this place as a relay and that they can produce something different through the mode of this program. I look forward to the program being also education-oriented so much so that it can perhaps alter the creative scene in Taiwan, which is now overly homogenized.

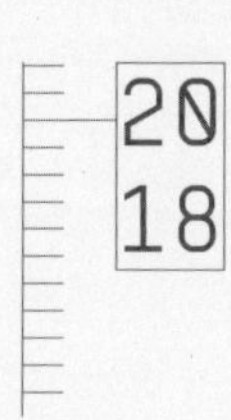

05.31-06.29 2018 CREATORS 進駐研發／創作計畫｜徵件

06.09 2018 CREATORS 徵件｜北部說明會

06.10 2018 CREATORS 徵件｜南部說明會

08.14 2018 CREATORS 進駐研發／創作計畫｜結果公布

09.14 悲傷ㄟ曼波—臺灣弄鏡文化的馬戲與錄像創作｜創作就像翻上一馬戲多元風貌的創作分享｜講談

09.19-23 Audiovisualizer in da house.｜Audiovisualizer in da House #1 光形成的結構｜表演

09.23 表演的政治性—惹內的陽台在台灣｜〈索多瑪的 120 分鐘—I feel〉逼近溺斃式的沉浸式體驗｜表演

09.27 A/V 衝刺班—實驗聲音影像推廣計畫｜A/V 衝刺班：音像藝術分享講座

09.30 「刷臉」時代的反統治鏈｜刷臉時代的反統治鏈—前導介紹與計畫說明會

10.05 表演的政治性—惹內的陽台在台灣｜「聊」・「育」・「戲」系列 1—什麼是「觀眾參與」？｜講談

10.06 悲傷ㄟ曼波—臺灣弄鏡文化的馬戲與錄像創作｜台灣喪葬科儀中的儀式表演｜講談

10.07-11.18 「Lands」殘片文化實驗計畫｜提著行李箱，拼出下一個風景｜工作坊

10.13 沉浸式逝去影像考古計畫｜3D Scanning Workshop II｜工作坊

10.14 表演的政治性—惹內的《陽台》在臺灣｜野孩子肢體劇場—《窯。臺》演員徵選

10.17-20 A/V 衝刺班—實驗聲音影像推廣計畫｜A/V 衝刺班：愛的波型三連發｜工作坊

10.20 ACCELERATOR 催落！— CREATORS 2018 開放工作室｜系列活動

10.24 A/V 衝刺班—實驗聲音影像推廣計畫｜A/V 衝刺班：演算世界／王者再起｜工作坊

10.28 回聲像—動態影像論壇與音像創作實驗計劃｜回聲像—動態影像論壇 Vol.1 什麼是動態影像？｜講談

11.23 悲傷ㄟ曼波—臺灣弄鏡文化的馬戲與錄像創作｜教你騎混種的野馬：探索馬戲創作的疆界｜講談

12.01 回聲像—動態影像論壇與音像創作實驗計劃｜回聲像—動態影像論壇 Vol.2 現場電影—以音像形式作為感官研究與實驗方法（如何實驗？）｜講談

12.08-23 譯譜者：在譜間轉譯的研究創作計劃｜開放的譯譜室｜工作坊

12.18 「刷臉」時代的反統治鏈｜刷臉時代的反統治鏈：數位統治術中的政治軟體｜講談

12.20 悲傷ㄟ曼波—臺灣弄鏡文化的馬戲與錄像創作｜劇場創作走向民間的痕跡｜講談

12.22-01.05　沉浸式逝去影像考古計畫 | 3D Scanning Workshop | 工作坊
12.28-29　空氣結構 lab—在島上呼吸 | We are Air-architects | 展覽

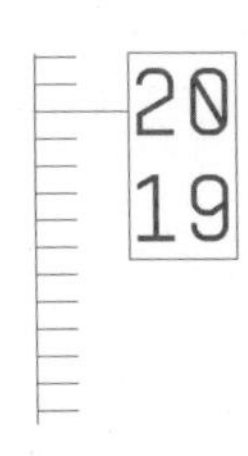

01.03　原住民族文化跨藝平台—藝術誌研發計畫 | 原住民表演藝術推廣平台 2015-2017 回顧推進策略 & 實務現場 | 講談

01.06-20　「Lands」殘片文化實驗計畫 | 殘片拼貼實驗工作坊

01.20　譯譜者：在譜間轉譯的研究創作計劃 | 譯譜者：2019 Lecture Performance | 表演

01.22-27　悲傷ㄟ曼波—臺灣弄鏡文化的馬戲與錄像創作 | 《悲傷ㄟ曼波》階段性展覽 / 座談 / 演出
表演的政治性—惹內的《陽台》在臺灣 | 惹內三部曲之二 《窯。臺》表演

01.26　ACCELERATOR 催落！— 2018 CREATORS 開放工作室 2019、徵件說明會 | 系列活動

01.30　「刷臉」時代的反統治鏈 | 臺灣當代藝術是否已經沒有合適的「文體」來詮釋當下的社會介面？ | 講談

02.02　Audiovisualizer in da house. | EBB VOL.4 Audiovisualizer in da 空總 | 表演

04.10　2019 CREATORS 創作 / 研發支持計畫 | 結果揭曉

05.27-10.31　當代城市採集—自己家的雜草茶 | 綠地認養：雜草志工招募

06.01　規訓的星系：戰後外省「離散」及「鎮壓」史調查創作計畫 | 規訓的星系 A：空軍「白色恐怖」的歷史圖像 | 講談

06.14-07.14　MASHUP all the CREATORS 搗亂了所有的創造者 | 展演

06.18-09.11　規訓的星系：戰後外省「離散」及「鎮壓」史調查創作計畫 | 離散與規訓、動盪的認同：1949 年後「外省」白色恐怖讀書會

06.21-07.21　當代城市採集—自己家的雜草茶「城市採集生活」徵文

06.23　規訓的星系：戰後外省「離散」及「鎮壓」史調查創作計畫 | 海軍「白色恐怖」的歷史圖像 | 講談

06.29　再現．抵抗．瓦解：一次重訪台灣同志污名史的邀請 | 我們的同運記憶 | 系列座談一

07.07　再現．抵抗．瓦解：一次重訪台灣同志污名史的邀請 | 我們的同運記憶 | 系列座談二
規訓的星系：戰後外省「離散」及「鎮壓」史調查創作計畫 | 規訓的星系 C：陸軍『白色恐怖』的歷史圖像 | 講談

07.14　當代城市採集—自己家的雜草茶 | 當代城市採集的探索與想像 | 座談會
規訓的星系：戰後外省「離散」及「鎮壓」史調查創作計畫 | 規訓的星系 D：軍醫的「白色恐怖」兼政治犯「屍」的處理 | 講談

07.20 ACCELERATOR 催落！—— CREATORS 2019 開放工作室 | 系列活動
07.21 規訓的星系：戰後外省「離散」及「鎮壓」史調查創作計畫 | 規訓的星系 E：「寶島夜船」與「反共義士」的白色恐怖 | 放映
08.02 規訓的星系：戰後外省「離散」及「鎮壓」史調查創作計畫 | 規訓的星系 F：「白色恐怖」下的「僑界」肅清 | 講談
08.18 城市塵埃—實踐街 1 號登月計畫 | 廢墟地圖——從非常廟與國家氣，談九〇年代空間佔領行動 | 講談
Ferment the City! | 共創塑膠再生魔毯工作坊
09.28 污痕結構學：城市記憶空間的建築文化學實驗 | 污痕結構學：建築的記憶與記憶的身體 | 講談
10.05 污痕結構學：城市記憶空間的建築文化學實驗 | 我們與監獄的距離：矯正建築、隔都（ghetto）與城市離散記憶 | 講談
10.05-06 污痕結構學：城市記憶空間的建築文化學實驗 | 污痕結構學：記憶組構工作坊
10.12 「延遲青春」座談 + 世界建構工作坊
10.18-20 社會運動交織而成的共同創作 | 畸零地 × 掀海風《苑裡好人》| 講談
10.19 城市塵埃—實踐街 1 號登月計畫 | 共享空間——從國有之地到想像之所 | 講談
Hidden Layer of City [臺北版] | 植入的音景工作坊
10.20 城市塵埃—實踐街 1 號登月計畫 | 缺席空間——重逢在月光樹影下的此曾在 | 講談
Ferment the City! | EM 菌解密實驗工作坊
10.26 ACCELERATOR 催落！—— CREATORS 2019 開放工作室 | 系列活動
11.02 無用便利店 | 便利の造型植栽工作坊
12.23 身為藝術工作者，我們如何組織自己？ | 身為藝術工作者，「共」的 n 百種方式？系列討論會二
12.29 Audiovisualizer in da house. | 三維梵現的音序 | 講談

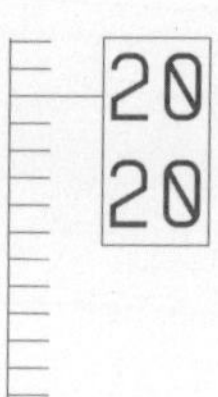

01.04-03.01 「2020 CREATORS 創作 / 研發支持計畫」徵件
04.30 2020 CREATORS 創作 / 研發支持計畫 | 結果揭曉
07.04-08.22 如果，家族旅行最終章：研究與創作實驗發展計畫 | 共學系列—不只是魅影：聲音和影像的卸除與穿越 | 講談
07.15-08.12 2020 CREATORS 系列講座
07.18 自然復刻—臺灣 AI 數位風景 | 共學系列：AI 謎音—演算法中的樂譜資訊 | 講談

08.09 負地理學：近未來的蓋亞實踐與藝術計畫｜火山，外星人與地球空洞說：一個域外科學的視角｜講談

09.16 糖蜜、酒精、健身工坊，是什麼使今日的生活變得如此不同｜空總盃：健身工作坊

09.19 ACCELERATOR 催落！—— CREATORS 2020 開放工作室

09.19-20 自然復刻—臺灣 AI 數位風景｜如果能和 AI 共譜一首樂曲？ AI 音樂共創工作坊

09.27 負地理學：近未來的蓋亞實踐與藝術計畫｜地震、火山與災難新聞：一個風險傳播的視角｜講談

10.04 負地理學：近未來的蓋亞實踐與藝術計畫｜飛碟，火山與外星生命：兼談東亞飛碟學研究的建構、聖域、地洞群與小矮人傳聞：一個現地調研與實踐的報告｜講談

10.11 負地理學：近未來的蓋亞實踐與藝術計畫｜地震、岩漿庫與火山生命：一個火山學的視角｜講談

11.01 負地理學：近未來的蓋亞實踐與藝術計畫｜火山，古文明與台灣島史：一個深層歷史的觀點｜講談

11.08 負地理學：近未來的蓋亞實踐與藝術計畫｜感應，能量點與火山地理磁場：兼談地外訊息接收｜講談

11.22 負地理學：近未來的蓋亞實踐與藝術計畫｜地震，海底山崩與北台灣的海底火山群：一個海嘯科學的觀點｜講談

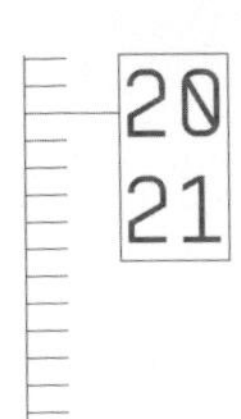

02.19-03.14 Signals 實驗通信：CREATORS 2020-21｜展演

02.19 如果，家族旅行最終章：研究與創作計畫｜家族旅行 No. 1—序｜表演
嚷嚷時間交換所｜淑枝與天球的黑膠分享會｜表演

02.20 綠頭鴨之死｜關於《綠頭鴨之死》計畫｜講談

03.06 投聲計畫｜投聲系列：校音一｜作品體驗
芥面：正體中文版 1.0｜芥面：新春吉利開運套餐｜講談

03.07 亞洲傳統音樂身體培訓交流計畫｜向傳統藝術靠近｜講談
投聲計畫｜投聲系列：校音一｜作品體驗
凝結歷史殘響｜回聲 / 反射 I｜表演

03.13 人工自然｜引之微觀｜表演

03.14 日常截面｜A Section of___｜表演

2018-2020 CREATORS 團隊計畫總表

年度	計畫名稱	計畫團隊
2018	Archive or Alive—劉守曜獨舞數位典藏	在地實驗
	空氣結構 lab—在島上呼吸	空氣結構 lab
	悲傷乀曼波」—臺灣弄鏡文化的馬戲與錄像創作	圓劇團
	回聲像—動態影像論壇與音像創作實驗計畫	ART SHELTER
	沉浸式逝去影像考古計畫	黃偉軒
	譯譜者：在譜間轉譯的研究創作計畫	謝杰廷
	Audiovisualizer in Da House.	燧人氏
	「Lands」殘片文化實驗計畫	劉時棟
	表演的政治性—惹內的《陽台》在臺灣	野孩子肢體劇場
	A/V 衝刺班—實驗聲音影像推廣計畫	噪流
	「刷臉」時代的反統治鏈	施懿珊
	身為藝術工作者，我們如何組織自己？	吳孟軒
	原住民族文化跨藝平台—藝術誌研發計畫	鄒欣寧
2019	Hidden Layer of City [臺北版]	張永達
	Relight—消逝與正消逝的臺北聚落光實驗計畫	莊知恆
	熒惑蟲計畫	僻室
	城市塵埃—實踐街 1 號登月計畫	郭奕臣、林怡秀
	當代城市採集—自己家的雜草茶	雜草稍慢 / 林芝宇
	C-LAB 園藝—打造一個永續的系統	王維薇、蔣慧仙
	規訓的星系：戰後外省「離散」及「鎮壓」史調查創作計畫	安魂工作隊
	再現．抵抗．瓦解：一次重訪臺灣同志污名史的邀請	沃時文化有限公司
	磁帶音樂記譜及模組研發計畫	林其蔚
	延遲青春	顧廣毅
	無用便利店	孫于甯、劉上萱
	發酵城市	施佩吟、蘇映塵
	社會運動交織而成的共同創作	畸零地創造股份有限公司
	污痕結構學：城市記憶空間的建築文化學實驗	李立鈞、吳耀庭、謝杰廷
	頹傾城市	張立人
2020	投聲計畫	吳秉聖
	自然復刻：臺灣 AI 數位風景	陳志建
	亞洲傳統音樂身體培訓交流計畫	江之翠劇團
	芥面：正體中文版 1.0	遠房親戚實驗室
	噥噥時間交換所	走路草農 / 藝團
	如果，家族旅行最終章：研究與創作實驗發展計畫	她的實驗室空間集
	採集者與版畫家 - 的確是存在於二十世紀	林傳凱（安魂工作隊）
	神的棲所 GiR	黃鼎云
	精神與靈魂的治理之術	施懿珊
	七個在海上的人：賽鴿	黃博志
	綠頭鴨之死	許哲瑜、陳琬尹
	負地理學：近未來的蓋亞實踐與藝術計畫	許博彥、盧均展、盧冠宏、浮
	糖蜜、酒精、健身工坊，是什麼使今日的生活變得如此不同，如此有魅力？	羅懿君
	凝結歷史殘響：臺灣洞穴的感知敘事	李慈湄
	人工自然—臺灣聲響實驗室沉浸式體驗研究與創作計畫	黃苓瑄
	日常截面	黃偉

陪伴觀察員	相關網頁/網站/社群媒體
王柏偉	www.etat.com/etat-news/2018archiveoralive/
王玨頎	clab.org.tw/creators/2018-r2/
邱坤良 / 吳思鋒	www.facebook.com/thunarcircus/
蔡家榛	issuu.com/loisxwang/docs/___-_______
林怡秀	cargocollective.com/HWH
鍾適芳	clab.org.tw/creators/2018-r6/
小樹（陳弘樹）	www.facebook.com/zuirens/
朱貽安	clab.org.tw/creators/2018-r8/
吳宜樺	www.facebook.com/lenfant.s.theatre/
馮馨	www.facebook.com/fluidnoise/
楊成瀚	clab.org.tw/creators/2018-r11/
王聖閎	clab.org.tw/creators/2018-r12/
周伶芝	clab.org.tw/creators/2018-r13/
郭昭蘭	www.changyungta.com
孫平	clab.org.tw/creators/2019-r2/
王嘉明	www.house-peace.com
姚瑞中	www.facebook.com/moonlandingnews
周序樺	www.facebook.com/weedroot
蕭有志	lab.garden/about
胡淑雯	www.facebook.com/安魂工作隊-620079101825098
張亦絢	cultime.co
張世倫	www.linchiwei.com
王冠婷	www.delayedyouth.org/
陳依秋	www.instagram.com/s.s.s.mart/
賴彥如	www.facebook.com/fermentthecity/
閻鴻亞	www.groundzerocreate.com
洪于翔	clab.org.tw/creators/2019-s7/
賴依欣	clab.org.tw/creators/2019-s4/
林強	www.pingshengwu.com
王柏偉	yi-studio.net
蔡淩蕙	www.facebook.com/gangatsui/
蔡宏賢	labotdrs.wordpress.com
林倉亙	www.facebook.com/wgagriculture
郭昭蘭	www.facebook.com/herlabspace
胡淑雯	www.facebook.com/安魂工作隊-620079101825098
孫以臻	clab.org.tw/creators/2020-s3/
吳祥賓	clab.org.tw/creators/2020-s4/
孫松榮	clab.org.tw/creators/2020-s5/
許家維	clab.org.tw/creators/2020-s6/
洪廣冀	spacestudio1219.wordpress.com
黃姍姍	yichunlo.wixsite.com/artist
凌天	www.litzi-mei.com/projects/caves/
臺灣聲響實驗室	clab.org.tw/creators/2020-so-1/
臺灣聲響實驗室	clab.org.tw/creators/2020-so-2/

完整 CREATORS 團隊介紹、相關活動請見

CREATORS | 臺灣當代文化實驗場

C-LAB CREATORS 臉書專頁

ACCELERATOR 催落！CREATORS 開放工作室 OPEN STUDIO

CREATORS 開放工作室是以團隊為中心，展現實驗計畫的階段性揭露。自 2018 以來，CREATORS 的開放工作室皆以「ACCELERATOR！催落！」為核心概念，試探當每一位 CREATORS 專注實驗其計畫之時，C-LAB 作為一個培育平台所能扮演的

時間	活動名稱
2018.10.20（六） 2019.01.26（六）	ACCELERATOR 催落！—CREATORS 2018 開放工作室

時間	活動名稱
2019.07.20（六） 2019.10.26（六）	ACCELERATOR 催落！—CREATORS 2019 開放工作室

時間	活動名稱
2020.09.19（六）	ACCELERATOR 催落！—CREATORS 2020 開放工作室

角色，在空間、補助及相關支援下，實驗計畫是否能產生更大的動能，創造不斷增生、不斷拓展的文化實驗力，陪伴與協助 CREATORS 直直催落去！

CREATORS 團隊	相關活動
黃偉軒、在地實驗、劉詩棟、施懿珊、圓劇團、鄒欣寧、吳孟軒、謝杰廷、野孩子肢體劇場、ART SHELTER、燧人氏、噪流、空氣結構 lab	開放工作室 分享會 After Party A/V衝刺班—實驗聲音影像推廣計畫丨A/V衝刺班-音像影響:從「98 E.M.P 電子音樂實習」談起丨講談 回聲像—動態影像論壇與音像創作實驗計畫丨回聲像 Back to the Cave 未來寓言丨放映 「刷臉」時代的反統治鏈丨刷臉時代的反統治鏈：科技敘事中的詞語清創丨講談 Archive or Alive：劉守曜獨舞數位典藏丨獨舞數位典藏方案分享丨放映 再現．抵抗．瓦解：一次重訪臺灣同志污名史的邀請丨我們的同運記憶丨系列座談三 Ferment the City!丨發酵產生的社會變化丨講談
張永達、莊知恆、僻室、郭奕臣 x 林怡秀、雜草消慢 / 林芝宇、王維薇 x 蔣慧仙、安魂工作隊、沃時文化有限公司、林其蔚、顧廣毅、孫于甯、劉上萱、張立人、施佩吟 x 蘇映塵、畸零地創造股份有限公司、李立鈞 x 吳耀庭 x 謝杰廷	開放工作室 催落市集 再現．抵抗．瓦解：一次重訪台灣同志污名史的邀請丨我們的同運記憶丨系列座談三 Ferment the City!丨發酵產生的社會變化丨講談 Ferment the City!丨秋日發酵下午茶丨工作坊 當代城市採集—自己家的雜草茶丨雜草小誌新書發表會 頹傾城市《FM100.8》播映座談會
吳秉聖、陳志建、江之翠劇場、遠房親戚實驗室、走路草農/藝團、她的實驗室空間集、安魂工作隊、黃鼎云、施懿珊、黃博志、許哲瑜 x 陳琬尹、引爆火山工程、羅懿君、李慈湄、黃偉、黃苓瑄	開放工作室 催落市集 她的實驗室空間集丨山穴的野狼眠夢，山豬的島嶼流亡丨演出 陳志建丨如果能和 AI 共譜一首樂曲？ AI 音樂共創工作坊

觀察報告 REVIEW

每年邀請不同背景的觀察員，以混合了深度報導及評論的書寫行動，跟隨各項CREATORS計畫的發展歷程，透過另一個觀察視角，深化文化實驗的過程意義，為每項計畫的推進、轉折與困頓，提出新的思辨。

年度	篇名
2018	擴張記憶的空間：再訪記憶與目光所及之處
	在地實驗「Archive or Alive——劉守曜獨舞數位典藏研發計畫」觀察員手記
	切身而做：劉時棟斷簡、殘片裡的美學探索及省思
	《窯．臺》：野孩子的惡之華與其超越性別的反父權批判
	跟悲傷的人一起跳舞：火，以及眾的平面
	「譯譜者」計畫的觀察記錄
	從籌辦演出轉往 Audiovisual Art 創作者的培育
	不存在的存在：「空氣結構 lab」駐村計畫
	藝術家的聚集，所為何事？側記吳孟軒與其關於「共」的機制思考
	軟體與主權的想像性變遷：論施懿珊 CREATORS 進駐計畫的可能和侷限
	我們要說什麼樣的故事？
	實驗．影像．對話：回聲像計畫觀察報告
2019	還能如何延續「那個空間」？郭奕臣、林怡秀「實踐街 1 號登月計劃」
	不好不壞不在場的民主：畸零地工作室的文資關懷策略
	光照亮了誰？莊知恆的光實驗計畫與未知的暗
	過去從未過去，「安魂工作隊」的調查行動與歷史時空再考掘
	光波粼粼中映現的意識餘像：張立人、成媛、芮蘭馨《FM100.8》
	一面以城市／歷史記憶為材料的百納旗，談李立鈞、吳耀庭、謝杰廷的「污痕結構學」
	超越現代性分工邏輯的身體機器，從「磁帶音樂」到《人體同步模組》
	數據生態中自然法則的無法則，談張永達的「Hidden Layer of City [臺北版]」
	許你一個無性的未來？顧廣毅《延遲青春》
	# 地球很危險：「僻室」登陸火星的廢物、拔克與災難
2020	施懿珊的資訊（意識）戰與黃鼎云的預言（治理）術
	走向民眾生活的藝術創作
	聲音場景的取樣、錄製與再建置
	臺灣藝術創作計畫，先有雞？還是先有蛋？有雞就一定有蛋嗎？
	不在此時：時差的潛勢
	2020 CREATORS「江之翠劇場」與「她的實驗室空間集」的觀察報告
	我的專業被你拿去跨界？來自生科人的文化實驗觀察報告

作者/ 觀察員	CREATORS 團隊
林怡秀 *	黃偉軒
王柏偉 *	在地實驗
朱貽安 *	劉時棟
吳宜樺 *	野孩子肢體劇場
吳思峰 *	圓劇團
鍾適芳 *	謝杰廷
馮馨 *	噪流
王萗頎 *	空氣結構 lab
王聖閎 *	吳孟軒
楊成翰 *	施懿珊
周伶芝 *	鄒欣寧
蔡家臻 *	ART SHELTER
黃鈴珺	郭奕臣、林怡秀
謝鎮逸	畸零地工作室
謝鎮逸	莊知恆
謝鎮逸	安魂工作隊
黃鈴珺	張立人、成媛、芮蘭馨
黃鈴珺	李立鈞、吳耀庭、謝杰廷
謝鎮逸	林其蔚
黃鈴珺	張永達
黃鈴珺	顧廣毅
謝鎮逸	僻室
羅倩	施懿珊、黃鼎云
蔡喻安	羅懿君、林傳凱（安魂工作隊）
馮馨	李慈湄、吳秉聖
王瑀	黃博志、走路草農 / 藝團
沈克諭	陳志建、引爆火山工程
謝鎮逸	江之翠劇場、她的實驗室空間集
王順德	遠房親戚實驗室、許哲瑜和陳琬尹

*- 陪伴觀察員
無標示 - 年度觀察員

完整觀察報告內容請見《CLABO 實驗波》

CREATORS 2018-2020 文化實驗三年索引
Three Years: CREATORS 2018-2020

指導單位 | 文化部
出版單位 | 財團法人臺灣生活美學基金會
董事長 | 李靜慧
執行長 | 謝翠玉
主編 | 游崴
執行編輯 | 林怡秀
文字 | 王柏偉、王莛頎、王順德、王瑀、王聖閎、朱貽安、吳宜樺、吳思鋒、沈克諭、周伶芝、林怡秀、游崴、馮馨、黃鈴珺、楊成翰、蔡家榛、蔡喻安、謝鎮逸、鍾適芳、羅倩（按姓氏筆畫排序）
校對 | 林怡秀、游崴、李樺、高慧倩、王萱、劉郁青
裝幀設計 | 理式意象設計
編排設計 | 李孟杰、温其綸、張采箖、顧白樺
翻譯 | 汪怡君、王家傑、黃亮融
插畫 | 楊絲早
印刷 | 崎威彩藝有限公司
出版日期 | 2022.05

特別感謝 | 王惠娟、吳宜臻、李玟億、周亞澄、林哲宇、侯昱寬、粘隶芸、陳雅柔、黃意芝、楊志雅、賴香伶、簡逸君、魏妏潔（按姓氏筆畫排序）

定價 | 新臺幣 900 元整
ISBN | 9786269607419

Supervisor | Ministry of Culture
Published by | Taiwan Living Arts Foundation
Chairperson | Jean LEE
Executive Director | HSIEH Tsui-Yu
Editor-in-Chief | YU Wei
Editor | LIN Yi-Hsiu
Text | WANG Po-Wei, WANG Ting-Chi, Sirius WANG, WANG Yu, WANG Sheng-Hung, CHU Yian, WU Yi-Hua, WU Sih-Fong, SHEN Ke-Yu, CHOW Ling-Chih, LIN Yi-Hsiu, YU Wei, FENG Hsin, Clytie HUANG, YANG Chen-Han, TSAI Jia-Zhen, TSAI Yu-An, Yizai SEAH, CHUNG She-Fong, LO Chien (in surname stroke order)
Proofreader | LIN Yi-Hsiu, YU Wei, LEE Hua, GAO Huei-Cian, WANG Hsuan, LIU Yu-Ching
Editorial Design | Idealform CO.
Layout Design | LI Meng-Chieh, WEN Chi-Lun, CHANG Tsai-Lin, KU Pai-Hua
Translator | Nicole WANG, Jack WANG, Alex HUANG
Illustrator | YANG Hsu-Han
Printer | Qiwei Color Arts Co.
Publishing Date | 2022.05

Special Thanks to | Cynthia WANG, WU Yi-Zhen, LEE Wen-Yi, ZHOU Ya-Chen, LIN Che-Yu, HO Yu-Kuan, NIEN Dai-Yun, CHEN Ya-Rou, Cécile HUANG, YANG Chih-Ya, LAI Hsiang-Ling, CHIEN Yi-Chun, WEI Wen-Chieh (in surname stroke order)

NTD 900
ISBN | 9786269607419

國家圖書館出版品預行編目（CIP）資料

CREATORS 2018-2020 文化實驗三年索引 = Three years : CREATORS 2018-2020 / 游崴主編. -- 臺北市 : 財團法人臺灣生活美學基金會, 2022.05

352 面 ; 14 x 20 公分

ISBN 978-626-96074-1-9（平裝）

1.CST: 藝術展覽 2.CST: 藝文活動 3.CST: 藝術行政

901.6 111007456

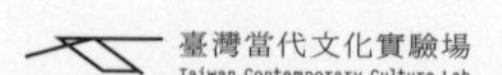

A 臺北市大安區建國南路一段 177 號
No. 177, Sec. 1, Jian-Guo S. Rd., Da'an Dist., Taipei

W clab.org.tw T +(886) 2-8773-5087 M info@clab.org.tw